FIRMAMENT

Simon Clark is a scientist, video producer, and online educator. Simon read Physics at St. Peter's College, Oxford before researching a PhD in atmospheric physics at the University of Exeter. During his studies he began creating YouTube videos about student life and his research, and has since accrued nearly 20 million views.

Firmament is his first book.

FIRMAMENT

The Hidden Science of Weather,
Climate Change, and
the Air That Surrounds Us

Simon Clark

HODDER

First published in Great Britain in 2022 by Hodder & Stoughton
An Hachette UK company

This paperback edition published in 2023

1

A CIP catalogue record for this title is available from the British Library

Paperback ISBN 9781529362312
eBook ISBN 9781529362299

Typeset in Bembo by Hewer Text UK Ltd, Edinburgh
Printed and bound in Great Britain by Clays Ltd, Elcograf S.p.A.

Hodder & Stoughton policy is to use papers that are natural, renewable
and recyclable products and made from wood grown in sustainable
forests. The logging and manufacturing processes are expected to
conform to the environmental regulations of the country of origin.

Hodder & Stoughton Ltd
Carmelite House
50 Victoria Embankment
London EC4Y 0DZ

www.hodder.co.uk

CONTENTS

Table of figures xi

Introduction: Giant xiii

Chapter 1: Idea 1

Chapter 2: Birth 21

Chapter 3: Wind 43

Chapter 4: Fields 55

Chapter 5: Trade 73

Chapter 6: Distance 93

Chapter 7: Forecast 109

Chapter 8: Vortex 131

Chapter 9: Change 147

Epilogue: Family 189

Endnotes 197

Acknowledgements 209

Glossary 211

Bibliography 215

Index 225

*The Lord hath his way in the Whirlwind, and in the Storm,
and the Clouds are the dust of his feet.*

(Nahum 1:3)
– quoted in The Storm, *1704*

Dedicated to Harry Backhouse, my Year 8 English teacher, who patiently sat with me on his lunchbreaks to edit my stories, and to all teachers around the world.
You make the future.

TABLE OF FIGURES

1 Aristotle's concept of the universe.
2 Carbon-12 and carbon-13.
3 How Ibn Muʿādh calculated the height of the atmosphere.
4 The vertical structure of the atmosphere, with layers defined by the vertical temperature gradient.
5 The electromagnetic spectrum.
6 A cumulonimbus cloud.
  © Tom Grundy / Alamy Stock Photo.
7 Halley's map of global wind patterns.
  © Darling Archive / Alamy Stock Photo.
8 A synoptic chart constructed by Fitzroy.
  Image provided by the National Meteorological Library and Archive – Met Office, UK.
9 A modern synoptic chart.
  © Crown Copyright [2015]. Information provided by the National Meteorological Library and Archive – Met Office, UK.
10 One formulation of the primitive equations.
11 The Keeling Curve.
12 F.H. Fritsch's engraving of exoplanetary solar systems.
  © British Library Board. All Rights Reserved /Bridgeman Images.

GIANT

This was a good spot. The air here smelled of redwood and rain, the gentle sound of the river lapping against his ears. Carefully, the young scientist set down his pack, full of the camping equipment you would expect on such a trip, but also including, more unusually, several large glass flasks. These were completely empty, not just of anything visible, but of any air whatsoever. Yes, he thought, this spot will do very nicely. Among the trees, away from other people. After setting up his tent and tidying his supplies, the scientist took one of these flasks and opened its valve. With a satisfying hiss, warm Californian air rushed into the evacuated glass flask, a tiny sample of the atmosphere here at Big Sur. The scientist resealed the flask, set it to one side and made a note of the date and time the sample was taken. He took a moment, listening to the sounds of the forest, letting the dappled sunlight play across his face. The next sample wasn't due to be taken for several hours. He had some time to enjoy the beauty here.

This was all a bit of fun for the scientist. Still fresh out of his PhD studies, the idea of taking his experiments on the road was frankly a good excuse to get out into the wilderness, seeing parts of the country he'd never seen before. He'd felt pressures from colleagues, superiors, and even the US government to pursue more lucrative studies, more useful to the national interest. Yet he felt drawn to these experiments.

They would lead to the discovery of a remarkable, shocking twist in the long history of our atmosphere. Never in thousands of years of study had one individual so fundamentally reshaped our relationship with the air around us.

And it all had to do with the tiny samples of the Earth's atmosphere held in those glass flasks.

The atmosphere is clearly of great importance to us all. It's the air we breathe. It enables life on Earth. Without this thin layer of gas, just a few tens of kilometres thick, our planet would be a frozen, lifeless rock. Yet few people can describe how it works. Or its composition, or its history in any detail. For such a vital part of our existence, the atmosphere receives remarkably little interest.

We are, of course, interested in the atmosphere *sometimes*. Particularly if you're British, the weather is a topic of endless fascination and frustration. Local variations in atmospheric temperature, pressure, and moisture are direct influences on our lives, the footprints of a giant, atmospheric entity. Yet most people's interest extends only to the footprints, not to the giant itself.

In fact, most people's broader knowledge of the atmosphere extends to just one topic. Climate change – and its associated uncertainties – has come to dominate the popular discourse around our planet's atmosphere: changes in the concentrations of carbon dioxide and other trace gases causing increases in temperature, rising sea levels, retreating glaciers, and more extreme storms. Or perhaps not. Perhaps the world is actually cooling due to changes in the Sun. Scientists seem unsure. The apparent uncertainty around climate change gives the impression that perhaps science still does not understand how our atmosphere works.

A viewer from outside the field could be forgiven for making this assumption: that science has yet to figure out how our planet's gaseous blanket works. After all, if we cannot seem to predict the weather with any accuracy, and we cannot work out if the climate is indeed changing, what *do* scientists actually understand? Yet the science of climate change and weather prediction are just parts of a modern, sprawling research field. Atmospheric science encompasses research in how layers of the atmosphere (yes, there are layers stacked on top of one another) talk to one another. Research in how the wind 50 km above the North Pole can be used to improve winter weather predictions, and save lives. Scientists for over 200 years have been probing, investigating, revealing the workings of our planet's atmosphere, and although there are still many unanswered questions, some of which we'll get to in this book, the giant has largely been understood. Science not only has an anatomical map of the giant – from its footprints, to the blood in its veins, to the hairs on the top of its head – but a textbook on its physiology. We understand what it is made of, where it came from, where it begins, and where it ends. We understand why it moves, where it chooses to flow. Perhaps most interestingly of all, we understand how it behaves as a vast entity, with events half a world away changing the state of the atmosphere and its footprints upon the surface.

Before I embarked upon my research career, I was entirely ignorant of this. My original scientific interests were a world away – I began my physics degree wanting to work in fusion power. But near the midpoint of that degree course I took a module on *geophysical fluid dynamics* – the physics of how the atmosphere and oceans move on the face of a rotating sphere, the Earth. The module opened my eyes to a whole new world of science, showing me it was possible to describe the behaviour of the entire planet using physical equations. Suddenly I wasn't considering infinitesimal point particles on a frictionless infinite plane, but real things – water and air! The physics of flowing fluids could be linked to the physics of temperature and heat, and those physics linked to those of the Sun and the Earth's orbit around it.

It was a glorious union of my different interests within physics with my passion for the natural world. I grew up reading Gerald Durrell and watching David Attenborough, getting lost in the woods of my local countryside, and playing around in the mud. All of this had been missing in my studies. To be able to combine these two worlds – of abstract physics and tangible nature – was a dream come true.

Along my path to eventually researching the interactions of the middle atmosphere and the surface, which will make a guest appearance in this book, I learned fascinating things. I learned how the temperature of the Pacific Ocean changes the length of winters in Europe. I learned how the lower atmosphere has a physical lid, preventing air from escaping any higher. I learned how above the polar night a storm the size of a continent spins at over a hundred miles per hour. This book is a journal of those discoveries. A condensed guidebook to our atmospheric giant. Starting from the earliest pioneers of atmospheric science, we'll meet a cast of characters from around the world – not only remarkable scientists, but also physics and phenomena that will open your eyes to how incredible the air we breathe is.

We'll begin by sketching out the anatomy of the atmosphere, introducing you to its layers, each with their own personalities and quirks. We'll look at the giant's blood – how and why air and moisture flow from one part of the globe to another, and the fascinating history of this understanding. Certain organs in the giant are particularly interesting, and we'll look at these in some detail – the great ribbon of air that forms the mid-latitude jet stream, the bulk of the colossal stratospheric polar vortex. Perhaps most mind-boggling of all is the way that the atmosphere exhibits *teleconnections*, the phenomenon of particular see-saw patterns influencing the giant's footprints on our lives even from the other side of the world, and we'll meet these in turn too.

As I've already said, most people are only aware of the atmosphere as an entity within the context of climate change. And when it is viewed in this way, our understanding of the atmosphere, as scientists, can seem incredibly flimsy. The main reason I wanted to write this book was to demonstrate that the science of climate change is only part of what we know about our atmosphere, and that it grew out of, and now nestles within, a mature field based on centuries of thought and experiment. It is a tree in a green forest of knowledge. The study of how carbon dioxide warms the planet's surface reaches back well over a century, and the dynamics of how it is mixed in the atmosphere have been understood for decades, coming from a millennium of observations. While there is much about climate change that is still being actively researched, its fundamentals are deeply rooted in the wood of atmospheric science.

The point of this book is to document these roots, to show how the vines of climate change mesh into the broader forest. With this perspective, it is my hope that you will come away with not only a greater appreciation of how wonderfully complex our atmosphere is, but also an understanding of how we reached our current level of knowledge. Perhaps most importantly, you will also come away fully aware of the significance of our young scientist in California, and the all-important samples of air he was taking.

CHAPTER 1

IDEA

It is 1862. James Glaisher (1809–1903) had just entered a hotel in Ludlow, in the rural English county of Shropshire. To say that he'd had a rough day would be an understatement. He'd travelled over twenty miles, walking the last seven cross-country. The previous fourteen miles he'd travelled had been through the skies, and he had been unconscious for some of it. He'd definitely earned a rest.

Glaisher and his associate Henry Coxwell (1819–1900) were aeronauts, literally 'air sailors': the brave, pioneering individuals who travelled by balloon. They were an experienced team, having undertaken dozens of experimental flights together. Coxwell was a natural adventurer, having been fascinated by balloons as a child, and by 1862 he had been a renowned aeronaut for quite some time, with over 400 flights under his belt. Meanwhile, Glaisher, a heavy-set family man with an impressive set of mutton chops, came to aerial adventuring late in life. Most days he was sat behind a desk as the head of the Department of Magnetism and Meteorology at the Royal Observatory at Greenwich, London. Previously, he had worked in the field and in the lab, for the British Trigonometric Survey and as a research assistant at the University of Cambridge respectively, but he increasingly found himself in the basket of a research balloon.[1] His previous experience taught him that in order to better understand the natural world, scientists needed more data, and more *accurate* data, and sometimes that meant getting his hands dirty and collecting it himself. Glaisher represented a turning point in how we interact with our atmosphere. This Victorian gentleman, scholar, and adventurer was part of an extraordinary century of activity that transmuted a superstitious, even supernatural, field of study into one of the modern pantheon of sciences.

He commissioned Coxwell to construct a huge new high-altitude balloon – the *Mammoth* – with the aim of exploring the atmosphere over 30,000 feet in altitude (around 9.1 km). Originally, this endeavour did not involve Glaisher – who was in his fifties – himself flying; instead the idea was for Coxwell to train several young meteorologists to accompany him on the flights. Glaisher was forced to step up to the plate, however, when one of the trainee meteorologists 'declined' to ascend on a training flight with Coxwell. He subsequently wrote: 'I found that in spite of myself I was pledged both in the eyes of the public and the British Association [for the Advancement of Science] to produce some results in return for the money expended. I therefore offered to make the observations myself.'[2]

This makes Glaisher's involvement seem rather reluctant, but biographer Richard Holmes characterises the senior scientist as delighted at the opportunity. After starting his career with years of field measurements, followed by a great expanse of time behind a desk, perhaps Glaisher longed for the excitement of the cutting edge of exploration. It is difficult to imagine the man who wrote 'do not the waves of the aerial ocean contain, within their nameless shores, a thousand discoveries?' *not* finding the idea of balloon exploration exciting. He certainly proved himself to be an efficient and meticulous instrumentalist, capable of making incredibly rapid and accurate measurements even under stress.

The flight of 5 September 1862 in the *Mammoth* proved to be the most incredible, and most dangerous, of his career. The pair set off from Wolverhampton that afternoon at around 1 p.m. Wolverhampton was used as a launch site for two simple reasons. First, it was home to a friendly municipal gas company, which provided the coal gas that lifted the balloon. Second, it's about as far from the coast as it is possible to be in England, and thus minimised the chance that the balloon would disastrously come down in the sea. Landing is naturally the most hazardous part of ballooning, and especially with expensive, fragile scientific equipment on board such as thermometers, barometers, and anemometers, it was imperative to secure the safety of both humans and apparatus. To this end, the pair used a padded crash box for the apparatus, designed to be as accessible and quick to stow as possible.* On this particular flight, however, the landing was the least of Coxwell and Glaisher's worries.

By 1:53 p.m. the pair had ascended to 29,000 feet (8,800 m), the temperature dropping below freezing and the sky above them a dark, Prussian blue. The flight seemed to be proceeding much like any of the other flights the pair had previously undertaken, but Coxwell and Glaisher were in fact in serious danger. As they ascended, the steady rotation of the basket had caused the release-valve line in the balloon – essential for decreasing altitude – to become twisted and tangled, leaving no way for the balloon to vent gas and slow down or reverse the aeronauts' ascent.

---

* While the colour of this box is lost to us, the balloon was arguably equipped with the forerunner of the black box found in modern aeroplanes.

Coxwell noticed this almost too late, and, eight kilometres above the Earth's surface, was forced to climb out of the basket and up into the balloon to fix the line. Unless something was done, the men would asphyxiate when they reached the upper atmosphere. At 29,000 feet the air was already so thin that breathing was difficult, and both men were rapidly starting to feel the effects of oxygen deprivation. They had perhaps minutes to live. In the basket, Glaisher felt his legs give way, his vision becoming increasingly blurry, and he found himself incapable of supporting the weight of his head. He faded from consciousness. Meanwhile Coxwell, also struggling from the lack of oxygen, was attempting to wrestle the line out of the balloon and into the basket. Still the balloon climbed higher and higher, now at a rate of a thousand feet a minute. The air was so cold that his hands were beginning to freeze solid and he repeatedly lost his grip, nearly falling out of the balloon. Eventually, miraculously, he pulled the line free.

However, his hands were by this point entirely frozen and incapable of gripping on to anything. He was unable to climb back down. After a few desperate attempts, Coxwell was able to manoeuvre himself using his elbows and dropped back into the basket. But the balloon was still climbing. With his hands now blackened and immobile with frostbite, he couldn't pull the release valve. With one last effort, he took the line in the crook of his elbow, gripped it with his teeth, and pulled hard.

His efforts were rewarded, and he heard the gas start to vent from the balloon above. Resting, panting for a few moments against the balloon basket, he crawled over to the unconscious Glaisher and attempted to rouse him. Glaisher later recalled that Coxwell, having cheated death by mere seconds, had – in a wonderful display of Victorian etiquette – implored him, 'Do try to take temperature and barometer observations, do try.' Glaisher replied, 'I have been insensible', to which Coxwell unflappably said, 'You have, and I too, very nearly.'

Freshly revived, Glaisher took up a pencil and started to take observations as soon as he was able (having fortified himself with some brandy, carried in the balloon basket for just such eventualities). By this time the pressure readings were already increasing, and the maximum altitude the pair reached could only be estimated based on the rate of ascent and descent before and after the crisis with the release valve. This was later calculated to be around 32,000 feet, though possibly as high as 37,000 feet (9.7 km and 11.3 km respectively), above the surface. It was an altitude record that would stand for nearly forty years.

The *Mammoth* touched down without incident in a large grassy field in Cold Weston, an abandoned medieval village outside Ludlow. Coxwell and Glaisher, wanting to report their remarkable journey, then walked the seven or so miles to the town's train station, but were stymied by a lack of trains that day. Glaisher telegrammed an account of their adventure, which would be front-page news the following morning, before heading to a hotel with Coxwell to get dinner.

Coxwell and Glaisher were unaware of it, but they had achieved a world first. They were almost certainly the first humans ever to leave the lower atmosphere. Had Glaisher maintained consciousness and watched over their instruments, he would have made a groundbreaking discovery: that the Earth's atmosphere was more complex and multi-layered than anyone suspected at the time. A repeat flight was out of the question, however, as the pair had barely escaped with their lives. In fact, to this day, the two men hold the record for the highest altitude unassisted by breathing apparatus. Advanced as *Mammoth* was, the technology of the time simply did not allow for human exploration of such a remote environment, even though it was only a few miles above the surface of the Earth.

As all scientists are, Coxwell and Glaisher were limited by the equipment at their disposal. Our understanding of the atmosphere, and indeed of all areas of science, is intrinsically linked to the development of technology, as well as how that technology is used. To understand why their flight was so significant – and why it would have rewritten the rulebook of the atmosphere had they kept an eye on their instruments – we first need to understand how Coxwell and Glaisher thought about the atmosphere, and what they used to measure it.

Even before the invention of agriculture, our atmosphere was a subject of intense interest and speculation for humans, as our food supply was directly dependent upon it. Hunters and gatherers provided for their communities at the mercy of the weather. Atmospheric conditions such as temperature, wind, and rainfall shaped the migration of animals and the foraging bounty of plants, eggs, and fungi. After the invention of agriculture, the earliest civilisations grew up around fertile, arable land. This made yearly harvests possible, but their yield was still dependent on the weather, which dominated the flow of life. It's no surprise that some of the earliest surviving fragments of writing contain references to the weather.[3]

These were mostly observations of phenomena and can't be considered scientific in the modern sense. Being of such significance, however, the study of the heavens – both earthly and celestial – formed a key component of ancient religions. As early as 3500 BCE, the religion of pre-dynastic Egypt prominently featured the sky and incorporated rainmaking rituals.[4] Early societies believed that atmospheric phenomena were under the control of the gods – famously the Ancient Greek sky god Zeus controlled lightning and thunder, while Ancient Egyptians believed in a god of air and wind, Shu.[5] Many also believed that these phenomena were connected to the heavens above them, specifically to the motion of certain heavenly objects. The Babylonians left extensive records of 'astrometeorology', written by astronomer-priests[6] and preserved on clay tablets. These contain copious observations of astronomical phenomena and even a theoretical framework describing them, including predictions such as 'when a cloud grows dark in heaven, a wind will blow' and 'when a dark halo surrounds the Moon, the month will bring rain or will gather clouds'.

Traditionally, the first steps towards science are considered to have taken place in Ancient Greece, though this may well have more to do with the abundant preservation and subsequent assimilation of Greek knowledge than with any primogeniture. The concepts of *science* and *scientist* didn't exist until the nineteenth century, however, before which the terms *natural philosophy* and *natural philosopher* were used instead to refer to the study of nature and to someone who studied it, respectively. Again – traditionally – the first natural philosopher is considered to be Thales of Miletus (624–545 BCE), who, as well as being credited with inventing the concept of a mathematical theorem and associated proof, was intensely interested in weather.[7] Right from the very beginnings of science itself, the atmosphere has been there.

Thales was familiar with the Babylonian system of astrometeorology, and made his own meteorological predictions. His approach to these predictions was significantly different from those of the astronomer-priests, however: they required no gods. His was a world in which nature obeyed entirely secular, rational laws. For example, Thales travelled the ancient Mediterranean extensively and was said to have visited Egypt to witness the annual flooding of the Nile. While the Egyptians explained the flooding as being the result of the god Hapi's arrival, controlled by the pharaoh, Thales gave a different, natural explanation. According to his theory, northerly winds prevented the Nile from flooding most of the year by pushing the river water upstream. In flooding season, however, this wind disappeared and so the river would burst out, unopposed, into the floodplain. While ultimately incorrect, this theory was entirely based on natural concepts and did not rely on mysterious gods pulling unseen, unknowable strings. It was the very beginning of what we call atmospheric science today.

While we cannot say for sure if the Ancient Greek states were the first to study and rationally theorise about the weather, we can say with absolute certainty that these were where *meteorology* as a field was born. The great philosopher Aristotle (384–322 BCE) coined the phrase from the Greek for 'the study of phenomena in the sky' in a treatise titled *Meteorologica*, published around 340 BCE.* Aristotle was arguably the most influential meteorologist of all time. His theories on the weather dominated textbooks in Western civilisation until the seventeenth century,[8] and – with his omnivorous mind, working in disciplines from physics to philosophy, botany to psychology – he was hailed as *the* authority for nearly two millennia. While details of his life are patchy, it's certainly clear that he was entranced by the natural world, taking meticulous observations of the plants and animals of his one-time island home of Lesbos. This obsession extended to phenomena of the sky, and his four-volume *Meteorologica* was the first known attempt to systematically describe and discuss the weather.[9]

---

* This ultimately named the modern field, via French polymath René Descartes' 1637 work *La Meteorologie*.

Aristotle borrowed from Eudoxus of Cnidus (c.390–c.337 BCE) the idea that the universe was divided into concentric spheres with the Earth at its centre. Around the Earth was a terrestrial sphere bounded by the orbit of the Moon. Above this terrestrial region was the celestial sphere. In modern terms we can think of this as dividing the known universe into our planet and its atmosphere, and the outer space beyond. This division, Aristotle argued, necessitated separate fields of science for each region: *astronomy* for the celestial sphere, and his new field, *meteorology*, for the terrestrial sphere beneath. This sphere, extending to the Moon, consisted of four elements – earth, water, air, and fire – as adapted from Empedocles (c.494–c.434 BCE), arranged in concentric layers. Earth lay at the bottom with water on top, which was then topped with air, which in turn was topped by fire.

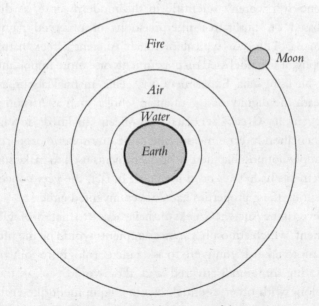

Figure 1: *Aristotle's concept of the universe.*

These layers were dynamic; for example, dry land stood above oceanic water, and fire often burned on the surface of the Earth. The four elements were in a constant process of interchange: when the heat of the Sun reached the Earth's surface, it mixed with the cool, moist water and formed a new warm and moist substance similar to air. Similarly, the Sun's heat mixed with the cold, dry earth to produce a substance that was warm and dry, similar to fire.

Despite being full of what to modern eyes are obviously flawed arguments, *Meteorologica* is a remarkable work. It summarises human knowledge of the atmosphere at the time, bringing together observations and prognostications from both Egyptian and Babylonian sources. It also represented a common theme of Ancient Greek natural philosophy: the evolution from supernatural to natural explanations. However, Aristotle's arguments still weren't scientific in the modern sense, as they were based on qualitative interpretations of observed natural phenomena. This was typical of much Ancient Greek natural philosophy: it was not based on experiments or – more importantly – falsifiable using data. Experiment was seen as manual labour, and so beneath the dignity of a gentleman scholar such as Aristotle.[10] And even if Greek scholars had been inclined towards experimentation, precise measurements of atmospheric properties were simply not possible then – the instruments used to make such measurements hadn't yet been invented. In fact, the very concept of measuring these properties hadn't been invented either!

But even if the Ancient Greeks did have access to meteorological equipment, which atmospheric measurements would be the most important to take? If you were to ask a meteorologist to join you in exploring some undiscovered land, they would pack in their bag – along with other essential scientific equipment like coffee – two crucial instruments: a *thermometer* and a *barometer*. The former measures the temperature of the surrounding air, while the latter measures the pressure this air exerts. Exactly what these concepts of temperature and pressure mean will be explained later, but for now it will suffice to say that they give scientists fundamental information about the state of the atmosphere at any particular location. By tracking how they change over time – along with other quantities such as humidity – scientists can not only classify the local climate, but, as we will see, also make predictions about its near future.

The meteorologist Sir William Napier Shaw once wrote that 'The invention of the barometer and the thermometer marks the dawn of the real study of the physics of the atmosphere, the quantitative study by which alone we are enabled to form any true conception of its structure.'[11] For our understanding of the atmospheric giant to evolve from that of antiquity to the one of pre-modern natural philosophy, these two instruments needed to be invented. As it would happen, both would first appear in Renaissance Italy.

The year is 1593. A hundred years have passed since Christopher Columbus limped back to port in Lisbon, bringing with him tales of the New World. Since then, immense wealth has been transferred to Europe, and some of this used to fund academics and scholars. One natural philosopher who received such funding, quite possibly the most famous of them all, was Galileo Galilei (1564–1642).

At this time, the study of the natural world was still dominated by Aristotle. For eighteen centuries his writings had been law on all that lay between us and the heavens – why the wind blows, where clouds form, and what causes rain. The *Meteorologica* was taught in universities across Europe and students learned that the Earth was surrounded by concentric spheres of water, air, and fire.

Galileo, however, was having none of this, and had a different idea of how natural philosophy should be done – using measurements, and mathematics.

Galileo had two advantages over the Ancient Greeks. First, he had access to far more advanced mathematical methods, including centuries of innovations from the Arab world and beyond, such as algebra and the concept of zero. Second, among other material improvements, he had *glass*. Glass, of course, dates back much further than the Renaissance, having been invented some four thousand years previously. But while Aristotle and other Ancient Greek philosophers had access to primitive glass, it was brittle, thick, and unusable for delicate items.[12] This would remain the case until the end of the thirteenth century, when a series of innovations was made by the glass makers of Venice. Experimenting with new compounds added to the silica that forms the body of glass, successive generations of glassblowers were able to make their products stronger and more malleable. They realised that by adding the burned ash of certain plants, their glass could be made perfectly clear, and far more resistant to attack by chemicals or sudden changes in temperature. This made the development of specialist scientific instruments possible – scientists could now isolate chemicals in their laboratories, or construct elaborate devices such as air pumps. With the combination of New World wealth, Arabic mathematics, and Venetian glass, Europe was poised to lay the groundwork for modern physics, chemistry, and biology.

Galileo would change atmospheric science forever. The first of his contributions came in 1593, when he introduced his *thermoscope*. Unlike a *thermometer* this device did not allow for measurements of temperature on an absolute scale such as Celsius or Fahrenheit – this would not be conceived for over a hundred years – but instead used relative measurements, by comparing whether a location became cooler or warmer over time, or whether it was cooler or warmer than another location. Like subsequent thermometers, however, Galileo's thermoscope worked on the principle of *thermal expansion*: material changing its volume when its temperature is increased. For most substances this means a very small increase in volume when temperature increases: for example, when water is heated from room temperature it very slightly swells in volume.* This occurs because increasing the temperature of a substance gives each component molecule more *kinetic energy*; i.e., each molecule in the substance wiggles around more vigorously, and to give themselves enough space they maintain a larger average separation with one another. In fact, this is what temperature *means* – it's simply the average kinetic energy of molecules in a substance.

Galileo used this effect by trapping water in a glass tube suspended in a pan of more water, sealed at the top with another glass vessel 'about the size of a hen's egg'. As the ambient temperature changed over time, so would the temperature of the water in the instrument and, due to the thermal expansion of its volume, the height it reached in the tube. Galileo proudly demonstrated his invention in a series of public lectures at the start of the seventeenth century, which does show how far we've come in the entertainment value of public scientific lectures. His instrument, however, was undeniably primitive. Over time it would be refined, first by enclosing the water in glass, preventing loss through evaporation, and later by replacing the water with other fluids – such as spirits and mercury – that experienced a greater change in

---

* It is interesting to note that this is not always the case. Negative thermal expansion is the phenomenon of substances contracting when their temperature is increased. This actually takes place with water between 0 and 4 degrees Celsius: when heated, it shrinks in volume.

volume when exposed to changes in temperature. In an interesting reversal of the patron/artist relationship, these improvements seem to have first been made by Galileo's patron, the Grand Duke of Tuscany Ferdinando II de' Medici.

The thermoscope would reach its modern recognisable form as the *thermometer* with the invention of an absolute temperature scale. This was a pan-European effort lasting a century, with early contributions from Anglo-Irish scientist Robert Boyle (1627–91), who attempted to set a fixed point for a temperature scale. For reasons known only to himself, he set this as the temperature at which aniseed oil (which he surrounded the bulb of his thermometer with) 'begins to curdle'.[13] This was not terribly successful as a reproducible method of defining a scale. Dutch mathematician Christiaan Huygens (1629–95) took things a step further by suggesting the use of two fixed points – the freezing and boiling points of water – and delineating degrees of separation between these fixed points to define 'a universal and determinate standard for heat and cold'.[14] Eventually, Polish-Lithuanian instrument maker Daniel Gabriel Fahrenheit (1686–1736) constructed the first reliable mercury thermometer in 1714, with his scale, still used today, based on three reference points: the equilibrium temperature of an ice/water/salt mix (that's zero degrees), the temperature at which ice forms on the surface of water (that's 32 degrees), and the temperature measured when 'the thermometer is placed in the mouth, or the arm-pit, of a healthy man' (that's 96 degrees).[15]

The temperature scale familiar to most people was invented by Swedish astronomer Anders Celsius (1701–44), who defined his scale as having one hundred degrees between the freezing point and the boiling point of distilled water. This is where the name 'centigrade' comes from, meaning 'hundred steps' in Latin. Originally, however, Celsius defined the boiling point to be at 0 degrees and the freezing point to be at 100 degrees: the opposite of our modern scale. The scale was inverted around the time of Celsius's early death at forty-two, allegedly by none other than the famous taxonomist Carl Linnaeus, also from Sweden, for use in his greenhouses.

With the mighty thermometer in hand, meteorologists were finally able to quantify the temperature of their surroundings. They could at last gather concrete data with which to develop, and test, theories about how the atmosphere functioned. Much of this data was trivially known – such as the fact that air grew colder at night, or became warmer as one moved closer to the equator – but some results were subtle. For instance, it was found that further from the Earth's surface, air temperature uniformly fell, though the rate of decrease was not the same everywhere. Temperatures also often fell in the hours before heavy rain arrived, and sometimes fell for seemingly no reason at all. As more data was collected, scientists began to notice patterns and construct new ideas about how our atmosphere was put together.

There was a problem, however – the thermoscope was in fact another instrument in disguise. Recall Galileo's original design for his thermometer: a glass tube topped with a bulb, suspended in a broad pan of water. The glass bulb at the top and the pan of water at the bottom isolate the water in the tube, preventing it from evaporating and changing the level in the tube. The pan of water had another unintended effect, however: it was also functioning as a barometer! Atmospheric pressure pushing down on the pan of water would influence the height of water in the tube; when the pressure was high, the water would reach higher. This would occur when there was a greater weight of atmosphere above. Why did the great Galileo not foresee this? Like most other scholars of the time, he didn't consider this possibility as he believed that air had no weight. This was a lingering remnant of the Greek system of the four elements, which taught that both air and fire possessed no weight.

If Galileo had any idea that this was a flawed assumption, he was out of time to change his mind. It was around then that he was courting controversy through his support of Copernicanism, the radical idea suggested by Nicolaus Copernicus (1473–1543) that the Earth in fact orbited the Sun, rather than the other way around.* This directly conflicted with the geocentric view of Aristotle, a view shared by the Catholic Church. After decades of decidedly bad behaviour towards the Church, Galileo was eventually found by the Inquisition to be 'vehemently suspect of heresy' and placed under house arrest in 1633.

In the last few years of his life, the frail and nearly blind Galileo employed a brilliant young mathematician named Evangelista Torricelli (1608–47) to be his assistant. Galileo had been impressed by Torricelli's work and had corresponded with him since his house arrest. Sadly, the two only worked together for a few months before Galileo's death in early 1642, but the self-described 'ardent Galilean' Torricelli strove to continue his master's research. In particular, against significant theological opposition, he developed the theory of 'indivisibles'[16] – an important precursor to calculus – and, more relevantly for our story, constructed the world's first true barometer.

In trying to improve Galileo's design, Torricelli filled a metre-long glass tube with mercury and carefully upended the tube into a basin, also filled with mercury. The level of metal in the tube, now vertical, fell gradually before stabilising one-quarter of the

* This is actually where we get the word 'revolutionary' from. Previously, it literally referred to one object revolving around another, but Copernicus's heliocentrism, with the Earth orbiting the Sun, was so radical that the term 'revolutionary' was used synonymously with monumental changes.

way down the length of the tube. However, after leaving his equipment for a few days, Torricelli noted that this level started to change. Some days the mercury in the tube fell rapidly, while on others it would slowly climb. Torricelli claimed that, with his changes to the instrument's design, these changes were caused by variations in atmospheric pressure, the weight of air overhead pressing down on the Earth's surface, rather than by changes in the temperature. Specifically, if atmospheric pressure increased, the level in the tube would grow higher, while if atmospheric pressure was low, the level in the tube would fall. He wrote, rather poetically: 'we live submerged at the bottom of an ocean of the element air, which by unquestioned experiments is known to have weight'.[17]

Such a claim flew in the face of prevailing wisdom that air possessed no weight. Extraordinary claims require extraordinary evidence, which Torricelli believed he had. But still, he needed allies to back him up. Fortunately, one of the other great developments of the scientific revolution came to his aid: the network of correspondents around Europe. The 'Republic of Letters' was an international, immaterial organisation of thinkers that formed in the sixteenth century and allowed for the rapid transmission of revolutionary ideas, much as the internet would at the turn of the millennium. Torricelli, being a gentleman scholar, was connected to this network in Italy via letter writing and organised visits from fellow scientists.

One such visit was made by the mathematician Marin Mersenne (1588–1648), who received a demonstration of Torricelli's barometer. Mersenne subsequently passed on what he had learned to French mathematician and polymath Blaise Pascal (1623–62). By 1648 Pascal had built his own barometer. He repeated Torricelli's experiment and came to the same conclusion: that air indeed had weight and exerted a pressure on the mercury in the basin. He further concluded that if air has weight, then the pressure exerted by the atmosphere should decrease as one climbed to higher altitudes, where the quantity – and therefore weight – of the atmosphere above was lower.

Conveniently enough, Pascal had a brother-in-law, Florin Perier, who lived near a mountain – the Puy de Dôme in central France. To test his hypothesis, Pascal tasked Perier with taking measurements with a barometer while climbing the mountain. Perier dutifully carried out this experiment and reported back to Pascal in a delightful letter that, indeed, the mercury in his tube fell lower and lower as he climbed to the mountain's peak. In addition to validating the theories of Torricelli and Pascal and dealing another blow to the supporters of Aristotelian physics, however, Perier's experiment suggested an interesting further conclusion.

Imagine that a mountain of some incredible height was to exist, and that you were able to persuade a long-suffering brother-in-law to climb it. As he clambers up the slopes, the weight of air above him would continually decrease. He would become increasingly out of breath as the atmosphere thinned around him. Eventually, if he were to climb high enough, and hold his breath long enough, there would be no weight of air remaining above him. The atmospheric pressure would have decreased to zero, and your (now completely breathless) brother-in-law would find himself in a vacuum. The atmosphere, it seemed, would simply peter out to nothingness at some unknown, great height.

Furthermore, it was well known by mountaineers such as Perier that the higher one climbed, the colder the surrounding air became: the summit of the Puy de Dôme was significantly colder than its base. It was a reasonable extrapolation to believe that as the atmosphere thinned with altitude, it would also become colder and colder. The Earth therefore must possess an atmosphere that was thick and warm close to the surface, and became thin and cold at great altitudes. This suggested the existence of some vast, freezing vacuum between the planets.

In modern parlance we would express the atmosphere's change in temperature with altitude as a *lapse rate*. Later experiments would determine that the lapse rate in the lowest few kilometres of the atmosphere – near the surface – is around -6 °C per kilometre. In other words, for every kilometre of altitude gained, the air temperature would decrease by around 6 °C. All observations showed a lapse rate somewhere around this value. Common academic thought was therefore that the atmosphere continued to decrease in temperature further from the Earth's surface until eventually the temperature of the vacuum of space was reached – presumed to be absolute zero: -273.15 °C – at around 50 km (31 miles) in altitude.

Coxwell and Glaisher may well have had this concern in mind in 1862, as they hurtled uncontrollably upwards in the *Mammoth*. Perhaps their principal fear was of freezing to death in the interplanetary void, never to return to Earth. Except that if Glaisher had been 'sensible' – to use his own words – he would have seen at his moment of maximum peril that something was deeply wrong: if he had looked into his frozen, hoary instruments, he would have spotted that while the pressure did continue to drop precipitously, the temperature was doing something entirely different. It wouldn't have been falling as the balloon ascended higher and higher: in fact, it would have been staying exactly the same.

CHAPTER 2

BIRTH

Travel to the Antarctic and dig straight down. Below the fresh snowfall you'll find – surprise, surprise – more snow. Beneath that there will be yet more snow. And more snow, and *more* snow. Every year more snow falls on the ice sheet and is in turn buried and compressed under the weight of subsequent snowfall.

Stop digging for a moment. In the walls of your snow pit you'll see a record of the previous years' snowfall, compressed into layers a few millimetres thick. You can distinguish one year from another by alternating winter and summer layers, much like the rings of a tree: in summer, less snow falls and the ice crystals are smaller due to the warmer temperature; while in winter, there is greater deposition and the ice crystals are larger. Digging down into the ice sheet therefore becomes a form of time travel: the further you dig, the older the snow becomes. Scientists have dug as deep as three kilometres down into Antarctica, drilling out cylindrical ice cores containing snow that fell as much as 2.7 million years ago.[1]

Amazingly, trapped inside these prehistoric layers of snow are tiny bubbles of air, sealed inside their miniature, icy prisons when the snow first fell and later entombed when subsequent snowfall compressed the layers of snow into ice. This trapped air forms a sample of the Earth's atmosphere as it was millions of years ago, and when scientists first analysed these little time capsules trapped in the ice cores, they found something incredible – the atmosphere then was not the same as it is today. In fact, the atmospheric giant has changed quite significantly over the course of its life.

Four and a half billion years ago, the Earth coalesced from the stellar nebula, the material in our solar system that was left over from the formation of the Sun. The first atmosphere of Earth was really just a part of this nebula, mostly hydrogen clinging to the surface of the planet, fresh out of the cosmic oven. As the Earth cooled and took shape, this hydrogen was gradually lost to space, and the planet was left exposed to the void. Here we find the origins of our current atmosphere.[2] Over several billion years, volcanic activity and the meteor impacts that were common on the young Earth spewed gases such as carbon dioxide, sulphides, and nitrogen into the sky. Then, as now, the majority of the atmosphere was nitrogen in diatomic form – N_2. This gas is common in the atmosphere as it's extremely stable both chemically and physically, not breaking apart even in the presence of solar radiation. Small amounts of nitrogen still persist in the rocks of the Earth, but the majority was outgassed to the atmosphere billions of years ago and remained there, inert.[3]

For the first billion years or so in the Earth's life, then, the atmosphere consisted of nitrogen mixed with smaller quantities of water vapour, carbon dioxide, and other trace elements. These are all still present in the atmosphere we know today, but there's a glaring exception. It's a gas that was only added to the mix after possibly the most momentous event in the history of our planet.

The earliest direct evidence of single-celled organisms dates to around 3.5 billion years ago,[4] in the form of microscopic fossils containing isotopes of carbon that can only be produced via biological processes.* To refresh your memory, *isotopes* are effectively different varieties or flavours of a given element. An *element* is a substance defined by the number of protons in its atomic nucleus – all carbon atoms, for example, contain six protons in their nuclei, while all uranium nuclei contain ninety-two protons. Different isotopes of an element all have the same number of protons in their nuclei, but different numbers of neutrons. Carbon, for example, comes primarily in two flavours: one with six protons and six neutrons in the nucleus, hence being called carbon-12; and another with six protons and seven neutrons in the nucleus, called carbon-13. When scientists analysed the 3.5-billion-year-old fossils, which, it must be said, really do look like rocks, they found the ratio of carbon-12 and carbon-13 present could only have been caused by biological processes. This process of using isotopes will come in useful again very shortly.

The earliest life forms were archaea, single-celled organisms distinct from bacteria, plants, or animals. The ancestors of plants were not far behind archaea, however, and within a hundred million years of the first fossils, life had developed the ability to photosynthesise. This seemingly innocuous ability to convert sunlight and carbon dioxide into energy, water, and oxygen would eventually have devastating consequences for the planet. Over the next billion years, life produced so much oxygen that the atmosphere started to change in composition. The oceans, previously

---

* It should be noted that there's even earlier evidence for life on Earth – certain rocks in Greenland were found to be 3.7 billion years old and containing hints of life in their formation, but this is less definitive than the 3.5-billion-year-old fossils. See A. Nutman, V. Bennett, C. Friend, M. van Kranendonk and A. Chivas, 'Rapid Emergence of Life Shown by Discovery of 3,700-million-year-old Microbial Structures', *Nature*, vol. 537, no. 7621 (2016), pp. 535–8.

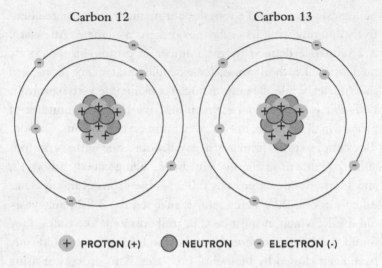

Figure 2: *Two isotopes of carbon: carbon-12 and carbon-13. Carbon-13 is heavier due to the presence of an additional neutron.*

anoxic, were flush with oxygen. This spelled disaster for the archaea, which had evolved to cope with an oxygen-free environment, and a mass extinction ravaged the planet in what is now called a variety of names, from the *Great Oxidation Event* to the *Oxygen Holocaust*.[5] This fundamental reshaping of life on Earth led directly to the huge variety of multicellular life we see today – certain organisms, in their attempt to survive the toxic, high-oxygen environment, clumped together to produce a safe haven for their genetic material.[6] Cells with nuclei and organelles like mitochondria were born. As I am a physicist by training, not a biologist, I will simply say here that the rest of the development of life is left as an exercise for the reader! By about half a billion years ago, the atmosphere was recognisably much as it is today – overwhelmingly nitrogen with a healthy dollop of oxygen, and a few trace gases such as water vapour, carbon dioxide, argon, and several others. Plants began to spread over the land, and the oceans teemed with complex animals. In a truly remarkable bit of detective work, scientists have been able to reconstruct what the Earth's climate was like even this far back in the past. Robert Berner (1935–2015), an American geologist at Yale University, constructed a statistical model[7] piecing together evidence from the weathering of certain rocks, carbon isotopes in other geological formations, and computer modelling. The details may be a little sparse, but this model, and others developed using similar methods, gives us a window over the past half billion years of the Earth's climate. It's safe to say that the atmospheric giant has experienced a few mood swings over that time: global temperatures in the Cambrian Period (541–485 million years ago [MYA]) were as much as 14 °C higher than the present day, while temperatures in the Permian Period (299 – 252 MYA) were maybe 3 °C colder than now.[8] A little closer to the present, different scientists have reconstructed the average temperatures of the Earth over the last 100 million years in much greater detail and found similar swings of more than 10 °C. This might seem difficult to believe, but the scientists' method of research was particularly ingenious! It involved oxygen isotopes.

The vast majority of oxygen on the planet is oxygen-16, containing eight protons and eight neutrons. Two other stable isotopes exist, however, and the more common one is oxygen-18, containing ten neutrons. Approximately 0.2 per cent of all oxygen in the world is oxygen-18, and a lot of this is bound up in H_2O – water molecules. Water molecules containing this heavier isotope of oxygen are themselves heavier,* slightly tweaking how they behave. Specifically, water molecules

---

* Not *heavy water* though! That is, water that contains deuterium – ^2H – rather than regular hydrogen. The two do have the same molecular weight, however.

containing oxygen-18 are more likely to be precipitated out of clouds as rainfall – they fall more easily due to their weight – and are also less likely to be evaporated in warm environments. These are truly miniscule differences, but even tiny changes add up to a great deal when averaged over an entire planet. Therefore, while sea water at the equator might contain a fair amount of oxygen-18, when this water is evaporated by equatorial heat, turning it into water vapour, there is proportionally less oxygen-18 in it. Also, when it's transported poleward, most of the oxygen-18 water – if it was even evaporated in the first place – will have been rained out before it ever reaches the poles. The water vapour that does reach the polar regions will fall as precipitation that contains almost exclusively oxygen-16 atoms. As a result, polar ice and glaciers fed by this rain, sleet, and snow will contain water molecules that are likewise almost exclusively made of oxygen-16, and very, very few of the heavier oxygen-18 atoms. As David Waltham notes in *Lucky Planet*: 'in the same way that evaporation and recondensation of fermented liquid concentrates the easily evaporated alcohol, so evaporation and condensation of seawater concentrates the easily evaporated "light" water. The Earth acts like a large whisky distillery.'

Therefore, the colder a planet is, the more 'light' water is trapped in the polar regions as ice, and the less oxygen-16 there is to go around – and so, the larger the overall ratio between oxygen-18 and oxygen-16 found in the warmer water areas. Or, in other words, the ratio of oxygen isotopes found in water molecules close to the equator can tell you roughly what the Earth's average temperature is. We know from subatomic physics how frequently the two isotopes of oxygen are formed, and so any deviation from this calculated, expected ratio of isotopes can be converted into a global temperature. More oxygen-18 than expected? The planet is cold (the oxygen-16 is locked in polar ice). Less oxygen-18? The planet is warm (there is little polar ice).

As neither isotope is radioactive, once oxygen of one flavour or another has been locked into a chemical structure such as a rock or fossil, it remains that isotope forever. This is perfect for scientists investigating past climates! All we need to do is find rock formations or fossils of a known age from the tropics and measure the ratio of oxygen isotopes found in them to estimate the temperature of the entire planet. Using this technique, we know that the Earth experienced sweltering heat 500 million years ago, being perhaps 14 °C warmer than at present, and slowly cooled to a minimum of a few degrees colder than modern temperatures around 20,000 years ago. Along the way it varied on much smaller timescales, getting hotter and colder over the course of a few thousand years. We will come to the reason for these cyclical variations much later in the book.

Since its formation, our atmosphere has clearly seen huge variations in both composition and temperature. The history sketched out in the past few pages is nothing more than a brief overview of the past few billion years, enough to show that we currently experience only the most recent few notes of a cosmic symphony that far precedes us and will continue long after we are gone. The atmosphere is never static, constantly evolving on short and long timescales. It is a planet-sized fluid, swirling and roiling, overturning and reconfiguring itself. But, for the last few hundred million years or so, it has been of approximately the same form. It has been made of the same molecules, and has extended to the same height above the surface of the Earth.

But just where is the edge of the atmosphere, exactly? How tall is it? This is a surprisingly difficult question to answer. The earliest estimation of the atmosphere's height actually came several hundred years before the invention of the barometer, by the Islamic scholar Ibn Muʿādh al-Jayyānī (989–1079). Little is known about him, and up until recently his work was incorrectly attributed to the incredible polymath Ibn Al-Haytham (Alhazen) (965–1040), best known for his treatise on optics and for emphasising the importance of experiments to evaluate hypotheses (several centuries before the scientific revolution in Europe). But it was in fact Ibn Muʿādh who calculated the first ever estimate of the height of our atmosphere, and his method was absolutely ingenious. It involved *twilight*.

Consider what happens when the Sun sets. While you are no longer directly illuminated by the Sun's rays, the sky is still suffused with a warm glow. This is because while the Sun is no longer visible to you at the surface, you would still be able to see it if you were higher up in the atmosphere, due to the curvature of the Earth. At the point of sunset, a thin sliver of Earth lies between you on its surface and the Sun, but slightly elevate yourself and your line of sight is no longer blocked. Of course, as the Sun continues to track across the sky, it will soon be obscured again by the Earth. To keep your line of sight, you must therefore elevate yourself a little higher again. This effect can be most clearly seen near mountains: the Sun sets marginally later when viewed from the top of a mountain, compared to when viewed from the bottom.

Ibn Muʿādh reasoned that twilight begins when the Sun sets, and ends when the entire sky is no longer illuminated by its rays[9] – or, in other words, when an observer at the top of the atmosphere can no longer see the Sun directly. Knowing how fast the Sun moves across the sky (about fifteen degrees per hour), Ibn Muʿādh timed how long it took for night to fall after sunset and applied his knowledge of spherical geometry to work out how far below the horizon the Sun must be when night truly fell, and thus the highest point above the surface that it could be seen by. Doing this, he estimated the atmosphere to be around 84 km (52 miles) tall.

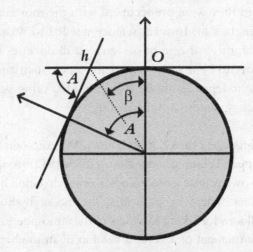

Figure 3: The geometrical arrangement of Ibn Muʿādh's twilight problem. The observer is located at point O and the zero-horizon illumination condition (when twilight has ended) is satisfied once the Earth has spun through angle A. The height of the atmosphere is determined using the half angle β=A/2 and basic trigonometry.

Especially considering the relatively simple (though ingenious) method used, this is not a bad estimate! However, Ibn Muʿādh had over-simplified several factors in his calculation. He had assumed that the atmosphere simply got colder and thinner with altitude. As we previously saw, the decreasing pressure, and hence density, of the atmosphere as altitude increased was shown in the seventeenth century by Blaise Pascal's keen brother-in-law. But this isn't the whole story. The atmosphere actually possesses a far more interesting vertical structure, with distinct layers sat on top of one another. Unfortunately for Ibn Mu'ādh, the first people in history who could have possibly challenged his simple atmospheric structure were Coxwell and Glaisher, but they were preoccupied with the mortal peril they found themselves in. Instead, science needed to wait until the twentieth century for one of the greatest discoveries in atmospheric history to be made. Remarkably, after such a long wait, it was made independently by two scientists, using very different approaches, almost simultaneously.

After a distinguished career at the French Meteorological Bureau, Léon Philippe Teisserenc de Bort (1855–1913) resigned and founded his own private meteorological research station in Trappes, near Versailles. There he pioneered the use of hydrogen-filled weather balloons for taking *soundings* of the atmosphere. A sounding is a measurement of a vertical column of atmosphere, using a term borrowed from mariners (depth sounding being the measurement of a depth of water, derived from the Old English *sund*, meaning swimming). Multiple measurements are taken above one location on the ground to determine how atmospheric characteristics – such as air pressure, temperature, water content, or a variety of other variables – change with height.

Soundings were originally done by scientists in balloons, such as Coxwell and Glaisher, or by attaching scientific instruments to kites, but, as we have seen, such methods suffered from severe limitations in the maximum altitude achievable. To overcome these limitations, Teisserenc de Bort attached lightweight instruments of his own design to kerosene-soaked paper balloons (the kerosene sealed any leaks at the joins) filled with lighter-than-air hydrogen. Upon release from the ground, the instruments would record changes in pressure and temperature as the balloons carried them aloft. As the apparatus climbed higher into the sky, the balloons would expand as the decreasing air pressure offered increasingly lower resistance to the hydrogen attempting to escape. Eventually, the air pressure became so low that the material of the balloons could no longer contain the hydrogen, and would burst. At this point a parachute would be deployed, and the instruments returned safely to the ground. This technique, pioneered by Teisserenc de Bort, has been refined over the years and continues to be used in meteorology right up to the present day. Approximately a thousand such balloons are still launched every day around the world by various meteorological agencies.

What Teisserenc de Bort found with his pioneering experiments was highly unexpected. As we learned in the last chapter with Coxwell and Glaisher, all previous soundings by crewed balloons and kites had observed a negative lapse rate – i.e. the higher they ascended into the atmosphere, the lower the temperatures they recorded. At first, Teisserenc de Bort's balloon flights recorded the same decrease; however, as they rose higher than any had before, the balloons observed this decrease only up to around 10 km in altitude, beyond which they began to record a lapse rate of zero. In other words, beyond this point, the temperatures they recorded remained *constant*!

Teisserenc de Bort originally believed that his instruments must be suffering from some error – perhaps, for example, being heated by the Sun in the high atmosphere – and so ran a few repeat experiments. Eventually, he published the results of 236 separate balloon launches, concluding that, indeed, the temperature in the atmosphere fell from a maximum at the surface to some minimum at around 10 km, and then remained constant as atmospheric pressure continued to fall with altitude. Furthermore, the altitude at which this 'isothermal layer' – as he called it

– began was higher above areas of high pressure and lower over areas of low pressure. Teisserenc de Bort suggested that our paradigm of the atmosphere as a single entity tapering off into the vacuum of space, as suggested by Pascal and Torricelli, needed to be revisited. He was reshaping our entire concept of the atmosphere by dividing it into layers, like a cake.

All of this was familiar to another scientist who was working on the same phenomenon at the same time, and in close correspondence with Teisserenc de Bort. The German meteorologist and engineer Richard Assmann (1845–1918), in contrast to Teisserenc de Bort, worked with a small team at his private research station and was a well-funded government meteorologist. While Teisserenc de Bort focused on completing a great number of balloon flights under many different meteorological conditions, Assmann meticulously improved his sounding instruments. As a result, while he made a mere handful of soundings, their measurements were considerably more accurate.

It was a classic case of quantity versus quality. But science is not a zero-sum game, and when quantity and quality are combined, progress can be made. Three days after Teisserenc de Bort announced his findings to the French Academy of Science, in collaboration with him Assmann announced his own to the German Academy of Science. He showed, using data collected from just six meticulously engineered soundings, a similar region with zero lapse rate – or a permanent *temperature inversion* zone. A temperature inversion is a localised, temporary 'inversion' of the typical temperature – i.e., a region where temperature increases with altitude. This can happen, for example, when freezing air rolls down mountain valleys and undercuts existing, warmer air.

Assmann had termed the new discovery as a permanent temperature inversion zone, while Teisserenc de Bort preferred 'isothermal layer'. Both names were equally valid descriptions of the observed phenomenon, yet neither name stuck. Our modern name for the region co-discovered by Teisserenc de Bort and Assmann – the *stratosphere* – was coined by Teisserenc de Bort some time later. The name comes from the Ancient Greek 'strata', or layers, and contrasts with what we now know to be the lowest layer of the atmosphere – the *troposphere*, taking its name from the Ancient Greek word for 'turn' or 'change', *tropos*. We'll discuss why these two names are so appropriate much later in the book.

As subsequent measurements would show, the stratosphere actually *increases* in temperature from the isothermal layer that Teisserenc de Bort discovered at 10 km to a maximum temperature at roughly 50 km in altitude. While the troposphere extends from the Earth's surface to around 10 km, the stratosphere extends from around 10 km to 50 km. The boundary between the two layers, the *tropopause* ('end of *tropos*'), varies in altitude, being higher near the equator and lower near the poles, but also, as Teisserenc de Bort discovered, varies in time. When the surface pressure is high, the tropopause rises in height. Equally, in a depression such as a storm or hurricane, when the surface pressure is low, the tropopause sinks down. The tropopause can be imagined as a thin manifold encompassing the world, constantly in motion, rising and falling with the movement of air masses in the troposphere and marking the edge of the atmosphere as we know it.

Yet that is not all – the atmosphere does not stop at the stratosphere! The stratosphere was only discovered with the development of balloon technology, lifting scientists and their instruments up to ever greater heights, but it had its limitations. Even today, the altitude record for a balloon is 53 km, set by Japanese researchers in 2002.[10] Balloons can only go so high before they reach a certain low-pressure limit, burst and tumble back to Earth.

Scientists needed a new way of travel, and in 1926 they got one: the liquid-fuelled rocket.

On 16 March 1926, in the snows of Auburn, Massachusetts, Robert Hutchings Goddard (1882–1945) launched the first rocket fuelled by gasoline and liquid oxygen. His singularly understated diary entry for the day reads, 'Tried rocket at 2.30. It rose 41 feet & went 184 feet, in 2.5 secs., after the lower half of the nozzle burned off. Brought materials to lab.'[11] There was no indication that his invention would revolutionise the twentieth century. Yet within twenty years, rockets had been co-opted for war. In another twenty they had propelled humans to space. Just forty-three years after Goddard's maiden flight, the technology he had pioneered landed humans on the Moon. Alongside these spectacular achievements, the rocket also revolutionised our understanding of the atmosphere, carrying scientific instruments above the stratosphere for the first time. The rocket that did so, however, was a weapon of war.

Goddard was only able to launch the first liquid-fuelled rocket 41 feet, or 12.5 metres, vertically. The utility of the technology was recognised almost immediately, however, and within a decade rockets had reached several kilometres in altitude. Particularly influential was work summarised in the PhD thesis 'Construction, Theoretical, and Experimental Solution to the Problem of the Liquid Propellant Rocket' submitted in 1934 by a young Wernher von Braun (1912–77). Von Braun was fascinated by rocketry from a young age. When he was just twelve, inspired by tales of rocket-propelled cars, he was taken into custody by local police for detonating a toy wagon festooned with fireworks in a crowded street (fortunately no one was injured).[12] Later he would become obsessed by the idea of space travel, and immersed himself in the study of physics and mathematics with the goal of one day visiting the Moon.

Circumstances of history, however, would lead him to work on developing weapons for the German military. The Nazi Party rose to power while von Braun was pursuing his doctoral studies, and his relationship with the Third Reich was complex and is still debated.[13] Whatever his relationship to the goals of the party, it was their war that enabled von Braun to advance rocketry to new heights. Working at Peenemünde on the Baltic Sea, he developed revolutionary technology for the *Vergeltungswaffe 2* – 'Retribution Weapon 2'. Better known as the V2, this was the first large-scale liquid-fuelled rocket in history. It was also a terror weapon, built by slave labour and designed to strike civilian targets at great distance without warning with a one-ton warhead. While it was of not much use as anything other than a psychological weapon – more people were killed in its construction than by its use[14] – it was of paramount importance in the history of science. Less than twenty years after Goddard's maiden flight of 41 feet, the first V2 test flight in 1942 reached an altitude of 84.5 km.[15]

Subsequent wartime flights would surpass this, but no measurements of the atmosphere were taken on any of these. They would only take place after the conclusion of the Second World War and the hasty transfer of rocketry supplies from Peenemünde to the United States. The accomplishments of the Nazi rocketry programme were much coveted by the Americans and the Soviet Union, and great efforts were made to secure both materiel and technical expertise in the closing days of the war. Fearing for their fate in Soviet hands, von Braun and his senior staff chose to surrender to the Americans instead, and delivered to them the most advanced rocketry programme in the world. After the war, the remaining V2 rockets were used for a variety of research projects, directly leading to both the US space programme and its ICBM arsenal.

Of greater relevance to our story was the rockets' use in exploring the upper atmosphere. In 1947 a V2 was launched from White Sands, New Mexico, reaching some 120 km in altitude. This was sufficient to detect not one but two further layers of the atmosphere for the first time.[16] Recall that in the stratosphere, the air temperature initially remains approximately constant, and then increases with altitude. What the V2 launch revealed was that above 50 km in altitude, the air temperature again stalls and remains approximately constant for a few kilometres before decreasing with altitude, much as in the troposphere. This continues up to around 80 km above the surface, above which the air temperature again increases with altitude. It thus marks two layers in the atmosphere: from 50 km to 80 km, and from 80 km upwards.

Apart from the instrument used – a slave-built rocket compared to balloons – what sets these discoveries apart is the manner in which they were reported. In a research paper in *Physical Review*, the results of the rocket flight is plotted on the same graph as a curve marked 'NACA estimated mean temperature', which they very neatly align with.[17] NACA was the National Advisory Committee for Aeronautics, the forerunner of NASA, and earlier in 1947 it had published a report indicating that the atmosphere was expected to cool with altitude before warming again, exactly as was observed. This was a beautiful example of science in action – sometimes the groundbreaking observations come first and are then explained by theory, and sometimes brilliant theoreticians make predictions that experimentalists subsequently verify. This particular bit of theorising was rather stunning, as it completely went against how science thought of the atmosphere just a quarter of a century earlier.

In 1923 it was still believed that air temperature was constant above the troposphere, as further measurements of the stratosphere had not been made. However, a study of how meteors burned up in the upper atmosphere indicated that the upper atmosphere must be much denser than was otherwise assumed, and thus much warmer (we will cover why in a subsequent chapter). It was suggested that temperature might actually increase with altitude.[18] Later, in 1934, a significant quantity of ozone in the atmosphere was first detected, indicating that a dense layer of the gas existed in the stratosphere, but not at higher levels.[19] Ozone was known to absorb UV radiation, which would warm up the air in the vicinity of the gas; this would explain the earlier meteor observations. However, as ozone concentrations were observed to fall with increasing altitude above the dense layer, the assumption was that air temperature above the then-observed atmosphere would fall again with altitude, as it did in the troposphere below. Around the same time, however, it became clear that the higher reaches of the atmosphere absorb high-frequency solar radiation more generally, and as a result could be *ionised*.

In the process of absorbing high-energy UV and X-ray radiation, gas in the upper atmosphere is ionised and greatly heated. It was theorised that the rate of absorption of high-energy radiation would increase above the stratosphere, and that there would be a region where this heating would produce a second layer in the atmosphere where temperature, again, increases with altitude.

The V2 flight of 1947 demonstrated that these predictions were correct, and that there were two additional layers to the atmosphere. These would later be named the *mesosphere* and the *thermosphere*, after their location between known layers (*meso-* being 'middle') and the property of air temperature rapidly increasing with height (*thermos* being 'heat') respectively. While the thermosphere – notable as the location of the spectacular aurora borealis and australis – has received a lot of research attention in the years since, I'm sorry to say that the mesosphere has been rather ignored. We certainly know more about it than when it was first discovered, though mostly what we have learned is that very little happens there – thus its rather unfortunate nickname among scientists as the 'ignorosphere'!

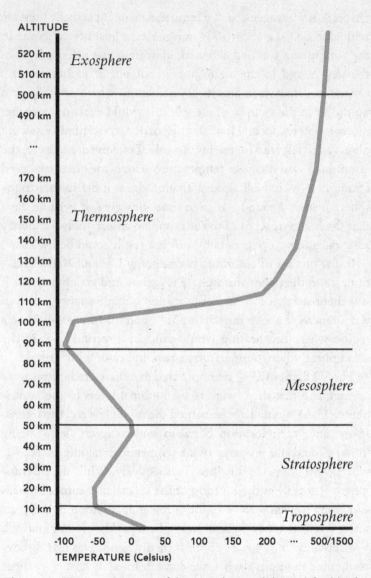

Figure 4: *The vertical structure of the atmosphere, with layers defined by the vertical temperature gradient (lapse rate).*

Even at the top of the stratosphere, the air pressure is less than a tenth of a per cent of its value at sea level. If we were to ascend beyond this point, into the mesosphere and then the thermosphere, the air pressure is so low that it ceases to have meaning. Do we classify the thermosphere as the top of the atmosphere? Can we even talk meaningfully of the top of the atmosphere? While it is divided into layers like a cake, the atmosphere does not have a pleasing dusting of icing sugar and thus a definite extent of any kind. If the atmosphere is made of molecules of gas, perhaps a definition could be the height at which all molecules disappear? However, even in the vacuum of space there is a vanishingly small density of atoms and molecules. How about the height at which the Earth is no longer the dominant gravitational pull on these molecules of gas? That would put the top of the atmosphere at around 1.5 million kilometres above the surface, which is much too far away to be meaningful.

In practice, there are a few definitions currently used to define the top of the atmosphere. The first – and most straightforward – is the Kármán line, named after aeronautics theorist Theodore von Kármán (1881–1963). This places the top of the atmosphere, and so the edge of space, at 100 km above the Earth, partly informed by how high an aircraft could fly in a straight line before travelling into space (i.e. not following the Earth's curvature in an orbit) and partly by one hundred being a nice round number. According to the Fédération Aéronautique Internationale, anything that takes place under the Kármán line is classed as aeronautics, while anything that takes place above it is *astronautics*. The V2 rocket was the first object to cross this line, and so was technically the first spaceship.

For spaceflight this is a sensible enough definition, but it isn't a *physical* one. Atmospheric scientists could argue that there are at least three other definitions for the top of the atmosphere. One is around the same height as the Kármán line, called the *turbopause*. This is the height below which the constituent gases of the atmosphere are well mixed, and above which they are instead stratified by how heavy the molecules are: the heaviest molecules in a layer at the bottom, with hydrogen in a layer on the very top. Another definition is the top of the thermosphere, at around 600 km in altitude, above which there are plenty of molecules – hydrogen and carbon dioxide mostly – that are still gravitationally attracted to the Earth, but there aren't enough of them to behave like a gas any more.

The final definition, really stretching what could be considered the atmosphere, is the edge of the final, outermost layer of the Earth's atmosphere, called the *exosphere*. This is nothing more than a scant few molecules still clinging to the Earth's gravity, barely interacting with one another at all. At around 10,000 km above the surface, the force that sunlight exerts on these rarefied atoms starts to outweigh the force of the Earth's gravity, and the planet loses its last tenuous grip on the molecules. Exactly where this occurs varies depending on the activity of the Sun, and on whether you are looking towards the Sun or in the lee of the Earth. It has recently been shown that these final strands of the Earth's atmosphere – known as the *geocorona* – can extend as far out as 600,000 km above the surface, meaning that the Moon occasionally passes through the Earth's atmosphere.

Now that we find ourselves thousands of kilometres above the Earth, the last few traces of atmospheric molecules floating by like strands of the giant's hair, let us look back at our planet. We see a green and blue marble, small to our eyes now, wreathed in the vanishingly thin envelope of its atmosphere. So far above the planet, the atmosphere is a wispy haze, a coat of gloss varnish on this globe hanging in space. While we may technically find ourselves just on the edge of the Earth's atmosphere here, some 99 per cent of the actual mass of the atmosphere lies within just 50 km of the surface. When compared to the huge size of the Earth, with a radius of some 6,400 km, the atmosphere is truly dwarfed.

CHAPTER 3

WIND

Weather is ancient. The English word 'weather' has its roots in the Old English word *weder*, which is believed to come nearly unchanged from Proto-Germanic, the hypothesised common language of north-western Europe spoken two and a half thousand years ago. Just like our obsession with the subject, 'weather' is an old word, perhaps stretching back in one form or another even further, to the Neolithic. Its meaning is well known to us: the state of the atmosphere around us. Weather is whether it is hot or cold, wet or dry, clear or cloudy. But 'weather' in its older forms originally had two connected meanings. The first was the one familiar to us. The second referred specifically to *wind*. In fact, the oldest lexical root we have for 'weather' is from the verb 'to blow', which shares a root with the modern English word 'wind'. In this sense, the root of *weather* is *wind*.

To me, wind really is the fundamental basis of weather. I would say that, though, given that my research of the Earth's atmosphere focused specifically on how air moves around the northern hemisphere. Without wind we wouldn't experience torrential rain, heatwaves, fog, thunder, storms, or any number of weather conditions. Wind is itself some weather, but additionally, wind enables *all* weather. The simple reason for this is that wind is our atmosphere's way of transporting material around the world, much as blood transports oxygen, nutrients, and waste around our bodies. The driving mechanism behind wind, then, is the heart of the atmospheric giant, pumping heat and moisture to all four corners of the globe, bringing weather with it.

But what is wind, really?

On a fundamental level, wind is the motion of air. When we experience wind, we are experiencing a current in the great ocean of air that is our atmosphere. In fact, mathematically, there is no difference between a current felt in the sea or a swimming pool and wind felt in the atmosphere. A gentle breeze is air in no great rush, a storm is a great stampede of air molecules racing from one point to another. On Earth, our wind speeds are rather modest – air typically sloshes around the planet at a rate of a few metres per second. The fastest wind speed ever recorded on the surface of the Earth was in Cyclone Olivia in Australia in 1996. A weather station clocked air briefly flowing at 113 metres per second (about 250 mph or 400 kph). This is peanuts compared to other planets, however. Neptune has the fastest winds in our solar system, perhaps three times as fast as Cyclone Olivia, and even that is dwarfed by winds detected around exoplanet HD 189733b, which were measured to be 2,414 metres per second (an astonishing 5,400 mph or 8,700 kph).[1]

Considering any planet's atmosphere as one entity, looking down on it from space, we say that it behaves as a *fluid*. This is a general term in physics that encompasses how liquids like water move, how gases like nitrogen and oxygen flow, and even how plasma in the heart of the Sun sloshes around. A fluid is a substance made of molecules and atoms, like solid objects, but made of molecules and atoms that only weakly interact with each other. We might imagine them to be like birds: one bird clearly flies as an individual, but put enough of them together and they behave as a flock. How this flock behaves often depends on its size; a small murder of crows, for example, may stick close together, but will act more or less like a group of distinct individuals. Put millions of starlings together, however, and they will behave as a *murmuration* – a flock that moves together as a single, colossal entity, flowing between buildings and parting like the Red Sea to evade a diving hawk. Watching a murmuration is like watching a sentient cloud, responding to changes in its environment – catching updrafts, chasing swarms of insects, fleeing from predators. Despite being made of individual birds, all obeying their own internal logic, the murmuration flies in a completely different way, greater than the sum of its parts.

In the same way, on a super zoomed-in level, wind on the Earth may appear as individually moving molecules of nitrogen and oxygen, but zoom out slightly and you'll see that they in fact flow together as cohesive groups. The atmosphere, then, behaves like an immense murmuration of starlings. Out of the simple motions of innumerable molecules weakly interacting with one another, beautifully complex behaviours emerge. These can be large scale, such as the circulation of the atmosphere around an entire hemisphere; medium scale, such as the pattern of winds around a storm; or small scale, such as the gusts of wind that flow around skyscrapers.

But what causes these behaviours? A murmuration of starlings will part to allow a hawk to plummet through, or follow a swarm of insects for their supper. It obeys the whims of the individual birds that make up the flock. Despite the analogy, molecules are not in fact living, breathing birds; so what *does* the atmosphere respond to, then? Can we still predict how the atmosphere might flow?

The answer to this is, of course, yes. But it took scholars and scientists centuries to pick apart what makes wind flow. It was only with this understanding in hand that our modern concept of the atmosphere, and the capability to predict future weather, became possible. As we will see in the coming chapters, the miracle of modern meteorology all flowed from an understanding *of* flow.

As we've already discussed, meteorology in one form or another has existed for thousands of years, evolving from astrometeorology based on interpretation of supernatural entities to a natural philosophy based on the reasoning of figures like Aristotle and then later Galileo and Newton. These approaches had something in common, however. They shared the belief that weather developed *in situ* – in other words, that the weather experienced at a location could be explained by applying the principles of your chosen system to phenomena *in that location*. Thousands of years ago, the weather in a locality was said to be determined by the action of a god like Hapi or Zeus at that location. Early natural philosophers meanwhile explained weather events using concepts like the exhalation of dry vapours, also in the locality of the events. Our understanding of the atmosphere took a great leap forward when it was realised that atmospheric flow – wind – could carry weather from one place to another.

The first inklings that weather could change in tandem across a large area are found in a book simply titled *The Storm*, written by then-political rabble-rouser Daniel Defoe (1660–1731), better later known as the author of *Robinson Crusoe*.[2] Defoe led a rather fascinating life that saw him in and out of prison, and at various times working as a spy, a journalist, a brick-maker, and as one of the first novelists in the English language. On the night of 26 November 1703, fresh out of one of his stints in prison, he witnessed a terrific storm. This would later – imaginatively – be named The Great Storm of 1703, and it was by all accounts a huge natural disaster. An estimated 8,000 lives were lost in shipwrecks, with around a fifth of the Royal Navy being destroyed, along with over a hundred deaths on land in England due to collapsing roofs and chimneys. The wind was so fierce that the lead roofing was blown off Westminster Abbey and one worse-for-wear ship was blown fifteen miles inland. We know about this event in such great detail because Defoe took it upon himself to collate reports of the destruction, placing newspaper adverts asking readers to submit their personal accounts of the disaster, which he then compiled and edited into a book. These reports didn't just come from England, however; others came in from across Western Europe, telling of windmills tearing themselves to pieces and livestock being blown out of their fields. Defoe surmised that these reports were all related to the same storm, one that had come from the west, passed over England, France, and Germany, and then continued out over the Baltic Sea, leaving destruction in its wake. This was no *in situ* weather event! Wind had carried the same atmospheric conditions over a large area.

Somewhat more peacefully, in 1743, American polymath Benjamin Franklin (1706-90) was attempting to observe a lunar eclipse in Philadelphia, but was thwarted by clouds.[3] His correspondents in Boston reported, however, that their view of the eclipse was good, with clouds only arriving an hour after the event ended. Franklin correctly realised that the clouds he was unfortunate enough to be under were the same as those later seen by his Bostonian friends. The clouds had been carried by a current of air from Philadelphia to Boston, weather again being transported from one place to another. Clearly then, in order to comprehend the weather at a given location it wasn't enough to examine the local causes – wind could carry weather that had developed elsewhere. There was no getting around it: an understanding of what caused wind was necessary to understand the weather!

This understanding emerged in the nineteenth century, starting with two hugely significant scientific papers that couldn't have been more different from one another. The first was written by American self-taught meteorologist William Redfield (1789–1857), a man with very little formal schooling but a clear talent for science.[4] In 1821 Redfield was out walking when he noticed that trees blown down by a storm had fallen in a geometric pattern – to be more specific, they had fallen in a spiral. This, he reasoned, must have been caused by the pattern of winds in the storm. Redfield went on to conduct an analysis of other storms that had similarly blown up the east coast of the United States from the Caribbean. He found that these storms all shared a similar pattern of winds, a pattern he described in a paper in 1831 as a 'progressive whirlwind' with a calm centre, rotating anticlockwise. Today we would call this pattern a *cyclone* (from the Ancient Greek *kúklos*, meaning 'circle'), a rotation of the fluid of the atmosphere around a point. Imagine a murmuration of starlings flying around a church spire. Those birds nearest to the spire must twist their bodies and change direction extremely quickly, while those further out leisurely soar as they make their way around the church. In the centre there may even be a few lucky birds who have perched on the stonework, sat perfectly still amid all the commotion. This is how the atmosphere behaves around a storm, rotating at great speed close to the centre and more slowly further out.* For reasons that will be explained later,

---

*   Of course, this is only true up to a point. Most storms have a calmer centre where the local pressure gradient is relatively small. Tropical cyclones such as hurricanes also have 'eyes' at their centres with barely any wind speed or cloud cover. These are caused by a counter-rotating current in the upper levels of the storm that forces some air to descend at the centre. This is a process that is still not fully understood! See J. Vigh, 'Formation of the Hurricane Eye', in *27th Conference on Hurricanes and Tropical Meteorology*, Monterey, American Meteorological Society, 2006.

this rotation is anticlockwise in the northern hemisphere, and clockwise in the southern hemisphere.

Redfield's storms were all 'progressive whirlwinds', but why? They must have all also shared a common cause. At this time, thanks to measurements by barometers, it was well known that the air pressure plummeted at the centre of a storm. The physics of fluid flow – well established by the mid-nineteenth century through the earlier work of mathematicians such as Daniel Bernoulli and Joseph-Louis Lagrange – indicated that where pressure was low, fluid would attempt to flow. Much as birds will fly from a desolate area to a verdant one with plentiful food, a fluid will attempt to flow into a region of low pressure. Naively then, given the low-pressure area in the centre of storms, we should expect air to rush in from the surrounding environment instead of rotating around the low-pressure area. But this was not what Redfield had observed! This apparent mistake was actually the cause of quite some controversy around Redfield and his so-called 'Circular Theory'. As it turned out, an additional force was needed to explain Redfield's observations, and it would be provided just a year later.

At this point in our story, we need only note that Redfield had bridged an important conceptual gap – through observations, he claimed that all storms had similar flow patterns in the atmosphere. What made a storm a storm? How intense its winds were, and specifically the direction they flowed in: an anticlockwise spiral. He connected this to the low pressure at the hearts of these storms, but couldn't work out precisely how this connection worked.

Here William Ferrel (1817–91) enters our story. Unlike so many other scientists in our story so far, he didn't come from a wealthy or respected family. Born to a family of farmers in rural

Pennsylvania, Ferrel was a shy boy who preferred to spend time away from friends, instead either working on the farm or devouring books.[5] He had a wide-eyed wonder about the natural world. Like many amateur scientists in this period, he was extremely motivated to learn more about nature, and overcame significant difficulties early in life to eventually pursue research. As a child he only received a formal education in two winters, taught with children of all ages in a single-room elementary school. This rudimentary education inspired Ferrel to learn more, however, and he made repeated, difficult journeys to towns in nearby Maryland to buy books on various topics in science. Over the years, Ferrel educated himself to the point where he himself could teach, eventually earning enough money to pay for tuition at college. After graduation he continued to absorb the drip of scientific information entering rural America, learning about the laws of physics that Newton (1643–1727) had derived and forming his own theories on the natural world. In particular he studied the theory proposed by Pierre-Simon Laplace (1749–1827) on the nature of tides, and (correctly) disagreed with Laplace's conclusion that the Moon and Sun were causing the Earth's rotation to accelerate. This disagreement formed the basis of the first scientific publication from the wide-eyed farm-boy at age thirty-six, in 1853.[6] Just three years later, Ferrel would establish an entirely new academic field.

'An Essay on the Winds and Currents of the Ocean' was published in a small, local periodical run by one of Ferrel's friends.[7] The paper, while not containing a single equation, is seen by some as the first in the impressive-sounding field of *geophysical fluid dynamics*, or GFD. While the name may sound like science fiction technobabble, at its heart the subject is really rather simple. So far, we've imagined a fluid like the atmosphere flowing like a murmuration of starlings. More specifically, this is like imagining a fluid on a static surface, such as water flowing on a table. But the Earth is of course not stationary! It rotates about its axis, and this massively impacts the way fluids behave on its surface. Allow that surface underneath a fluid to rotate in some way and a whole heap of complications arise. It's rather like a murmuration flying through a storm – birds might try to fly in one direction and find themselves diverted in another, the whole flock having their flight affected by the strong currents in the air. It becomes impossible to fly in certain ways: attempt to fly into the storm and the birds remain stationary, attempt to fly perpendicular to the wind and the birds instead follow a diagonal path. The faster the rotation of the surface the fluid rests on, the stronger the storm our birds must fly in – impacting on the ways the murmuration can fly more and more. This is the case for a two-dimensional flow, such as water on top of a rotating table. A *three*-dimensional rotating flow, such as an atmosphere flowing over the surface of a rotating sphere, is similar but more complicated again. The mathematics describing this deceptively complex impact on the flow of a fluid are really rather tricky, and are at the beautiful heart of geophysical fluid dynamics. Ferrel's 1856 paper was the first, tentative step in GFD, using verbal reasoning and the dynamics known at the time to deduce the existence of a new feature of the atmosphere.

It was followed up by a truly giant leap in 1858, when Ferrel published 'The Influence of the Earth's Rotation Upon the Relative Motion of Bodies Near its Surface'.[8] This tiny paper, just four pages long, is a true milestone in meteorology, crammed with elegant mathematics that seem to burst out of the margins. It was the first time that someone had combined all the pieces necessary to predict how air behaves on the surface of the Earth. Using data collated by mathematician and fellow meteorologist James Henry Coffin (1806–73), it combined Newton's second law (that the force acting on an object is equal to its mass multiplied by the acceleration it experiences, or F=mA) with rotational mechanics necessary to calculate the deflection a parcel of air experiences as it moves around a rotating sphere. Ferrel the farm-boy had put numbers to the flow of the atmosphere, and in so doing explained Redfield's 'Circular Theory': why air rotates around the low-pressure heart of a storm. Air rushes to occupy an area of low atmospheric pressure, yes, but in doing this it is deflected by the rotation of the Earth. This is known as the *Coriolis deflection*, and we'll talk much more about it in later chapters. More generally though, Ferrel had derived equations that roughly described how winds formed *everywhere*. If you knew the distribution of temperature and pressure in your area, even from just a few measurements, you could use Ferrel's equations to calculate what winds will blow as a result!

When I was an undergraduate, I studied a module on, among other things, how fluids flow. While sat in my college library one afternoon working out tutorial problems with paper and pencil, I utterly fell in love with the subject. There was one problem in particular that sealed the deal. A fluid was placed between two plates and one of these plates heated by a certain application of energy. How did the fluid between the plates respond? First, I applied the simple equations describing how the plate would warm as heat was transferred. Then, while applying the equations describing how the fluid would accelerate in response to this change in temperature, I was struck by what I was doing. With a mere handful of mathematics and a couple of pieces of information, I was able to describe the behaviour of a continuous, gooey fluid stuck between two planes. How its starlings were more free to fly near one plate than the other. In a physics degree you spend much of your time studying ridiculous things like the motion of a pendulum with no friction, or how a perfectly spherical chicken would roast in an oven, or the way two infinitely small electric charges repel one another. But here! Here was something tangible, something real. This was no infinitely small particle on an infinite, frictionless plane. I could immediately imagine this hypothetical scene taking place in front of me, and, more than that, I could see how it could be generalised. This particular question was about some oil-like liquid. But the Earth's atmosphere also behaved as a fluid, variably heated from below by the land and ocean. The exact same equations I had jotted down with my pencil could be used to calculate how the atmosphere would flow, produce wind, make weather. If I had information on the characteristics of the atmosphere and the temperature of the ground underneath, I could predict what weather would take place tomorrow! Of course, I could only dream at the time of being able to work out all the extra complexities of fluid flow taking place on a rotating sphere, making it truly applicable to the Earth's atmosphere. Ferrel was the one who had taken this conceptual leap, worked out the details, and made it a reality. All meteorology since owes him a great debt.

I would be remiss not to mention that such understanding was also empirically worked out – entirely independently – a little later by the spectacularly named Dutch scientist Christophorus Henricus Diedericus Buys Ballot (1817–90), though he would subsequently acknowledge Ferrel's primogeniture. Buys Ballot published his results in 1857 and had the distinction of validating his results by comparing wind data collected along the Dutch coast to barometer readings taken at the same time.[9] These showed that the equations describing the wind around areas of low pressure first derived by Ferrel and later by Buys Ballot were born out in real-world data. For this reason, despite him not being the first, history chose to remember Buys Ballot. What we call 'Buys Ballot's law' is still taught in military academies today:

in the northern hemisphere, if you turn your back to the wind, the atmospheric pressure is low to your left and high to your right. This statement, which may be familiar to you, is simply a very condensed form of the physical equations Ferrel and Buys Ballot derived. When put so simply, it's difficult to understand why it took so long for science to describe winds like this. Yet it took the development of the necessary equipment (the barometer) as well as necessary observations (via Redfield) and the necessary theory (via Ferrel and Buys Ballot) to piece this understanding together.

Buys Ballot's law has led to Ferrel being rather unfairly forgotten. He was a truly remarkable figure in the history of science, going from farm-boy to corresponding with the finest scientists of his day. His work lies at the heart of modern meteorology, and underpins our discussion in the coming several chapters: how wind brings weather, and shapes our climate.

CHAPTER 4

FIELDS

This will seem very strange, but stay with me.

Imagine a room, say your own bedroom. Fill it with as many cats, dogs, and birds as you can possibly find and lock the door. Pandemonium ensues. The dogs are having a great time, happy to have made so many new friends, while the cats are frantically running rings around them, trying to find a safe place away from the big brutes. The birds aren't bothered by the dogs, but instinctively fly away from the cats, flocking around the room, searching for a safe place to sit. For the first few minutes, the room appears to be in absolute chaos, a blur of fur and feathers. Eventually, however, things settle down. After much barking and bumsniffing, the dogs establish a pecking order in one side of the room. The cats position themselves sporadically throughout the room, many under the bed, keeping a wary eye on all the other animals. Some are looking upwards, tails twitching, at the birds, who have largely perched on the light fixtures. Other, bolder birds have settled on the back of the largest dog, foraging for food in the fur of a very patient golden retriever. Things have come to an equilibrium. Yet if this equilibrium is disturbed, if say the golden retriever has enough of being a bird buffet and gets up, there are knock-on effects for the other animals. The cats closest to the golden retriever scatter, fearful of what this terrifying beast could do next, spreading panic in the other cats. The birds on the dog's back take flight, causing some of the hungrier cats to give chase, which then disturbs more of the birds into flight. Some of these birds grab the attention of the dogs, who give chase, barking, scattering the cats even further. In reality, a true equilibrium between the animals would never be reached, with each small disturbance cascading into changes in the dogs, the cats, and the birds. The three types of animals would be forever dancing around one another.

A physicist would describe our pandemonium room as a *domain*, a stage, a physical volume, that action can take place on. Each of our types of animals – cats, dogs, birds – could then be described as a *field* within this domain. The cat field, for example, tells you where cats are within the room. In this approach, each square metre of the floor has a number associated with it, which is the number of cats within that square metre. All of these numbers together, describing the entire room, is the cat field. If we wanted to be fancy, we could summarise all this information in a mathematical object represented by a Greek letter, like γ. Equally, however, we have a dog field and a bird field, also defined on exactly the same domain. These fields could be represented by some other letters, such as δ and β. Each square metre of the room therefore has several numbers associated with it – the value of the cat field, the value of the dog field, and the value of the bird field; being the numbers of cats, dogs, and birds respectively over that square metre.

These fields are not static, however. As time moves forward, the number of cats under the bed, say, will change. Initially, there were no cats hiding under the bed frame, but as more and more moggies realise it is a safe haven, the value of the cat field under the bed, initially zero, grows larger and larger. Until, of course, something happens, like a dog sticking its snout under the bed, at which point the cats scatter, and the value of the cat field under the bed plummets back to zero. The dog field, meanwhile, is largely unchanged. The dog field near the bed had one dog in it before, and will have one dog in it after.

In case I've lost you along the way, let me just briefly restate what I'm getting at. Our room is a domain, in which we can define multiple fields on top of one another, each representing a different type of animal. The values of these fields correspond to how many of that type of animal there are in a certain part of the room. The interactions between animals show up as changes in our three fields over time.

Mathematically we would say that our fields *interact* with one another. In fact, we could even define some equation that links the values of the cat field, the dog field, and the bird field. This would be very useful if, for example, we didn't know anything about the cat field and yet had all the information about the dog field and the bird field. The equation would tell us, given some dog field and some bird field, what the cat field must be as some combination of the other two fields. If we wanted to use our fancy notation, we could even write it as something like:

$$\gamma = \delta \times \beta$$

This equation tells us that the number of cats at a location (γ) is equal to the number of dogs at that location (δ) multiplied by the number of birds at that location (β). Linking all of the *dynamic fields* (fields that change over time) together, we would refer to our equation $\gamma = \delta \times \beta$ as the *equation of state* for the room. An equation of state links together all of the fields that defines how a system behaves – in our case, the fields representing cats, dogs, and birds.

Clearly, however, this particular equation isn't very good! An accurate equation of state would almost certainly be more complex than simply multiplying the two other fields together to calculate the remaining field, perhaps instead saying that the cat field depends on the rate of change of the bird field, or how the dog field changes throughout the room (where there are the fewest dogs, you find the most cats). Piecing together the actual equation linking our three fields would involve either studying the psychology of cats and dogs and birds, or watching the room evolve over time and finding an equation that fits the behaviour we observed. While it is complicated, and describing seemingly chaotic behaviour, it would be possible to eventually derive the equation of state of cats, dogs, and birds.

At this point you may well be wondering what on earth does this have to do with the atmosphere?

Well.

Replace the room with our planet. Instead of a dog field, now consider a *temperature field* – specifically, how the air temperature varies throughout the atmosphere. Instead of a cat field, consider a *pressure field*, how air pressure varies throughout the atmosphere. Each location in the atmosphere therefore has two numbers associated with it: the pressure and the temperature at that location. Considering the pressure and temperature through the entire atmosphere defines the fields for these respective quantities. Previously, we also discussed how these fields interact, with changes in one field impacting on the other. So it is with the atmosphere also! In fact, we can define a very simple equation of state for the atmosphere:

$$p = \rho \times R_s \times T$$

The air pressure at a location (p) is equal to the temperature at that location (T) multiplied by two other factors. One of these is just a number, R_s, which is a physical constant,* and the other is the air density at that location (ρ). This is another dynamical field, equivalent to the bird field in our room analogy, that represents the mass of air in a cubic metre at a location.

This equation of state, which we also refer to as the *ideal gas law*, is a nexus of information in atmospheric science.[1] Much of the study of atmospheric physics is really the study of how fields like temperature and pressure, as well as others that we haven't even mentioned such as moisture content and aerosol density, interact with one another and change over time. The way that pressure and temperature and density and myriad other factors interact may seem chaotic and near-random, but their interactions can be condensed down to one equation: the equation of state. The equation of state is an indispensable part of the theoretical toolkit needed to investigate the atmosphere, acting almost as a universal translator of information. There are many occasions in which we will have some information about the atmosphere, such as the air temperature and the air pressure, and want to know something else, such as the air density. Simply apply the ideal gas equation, and away you go. I've gone into such detail, introducing it here not only because it represents how physicists think about the atmosphere in terms of interacting fields, but also because of its universality in practical studies of the atmosphere. Its importance really can't be overstated.

---

* This is the specific ideal gas constant for dry air, approximately 287 J kg^{-1} K^{-1}. If you are familiar with the ideal gas law $pV = nRT$, you may recognise the gas constant, but note that this is the *specific* gas constant, i.e., the gas constant divided by the molar mass of dry air. Of course, air in the atmosphere isn't necessarily dry! We can actually account for this very easily in the equation of state by changing from considering temperature to considering *virtual temperature*, which includes information on water vapour.

As it turns out, if we keep track of how air temperature in particular changes throughout the atmosphere, we can – via the equation of state – get most of the way towards understanding how air pressure changes. And, as we learned in the last chapter, variations in air pressure are the primary cause for the atmosphere to flow. So, wind is forced by air pressure, which is in turn forced by air temperature. William Ferrel derived the equations describing the first link in the chain, describing how pressure causes wind, and we have just ourselves derived the equation describing the second link in the chain: how pressure is affected by temperature.

We now find ourselves tantalisingly close to understanding the root cause of weather in the original meaning of the word! How the atmosphere behaves, and specifically how the atmosphere moves. The question remains then: how and why does air temperature change?

The answer hangs above us, blazing in the sky.

The Sun is a stellar forge, industriously converting hydrogen into helium. While this sounds grand and terribly complex, really our star is nothing more than an enormous collection of gas. So much gas in one place, in fact, that the extreme pressure and temperature at its centre are enough to ignite the process of nuclear fusion, where hydrogen atoms are squeezed together to form helium atoms. Just like a forge here on Earth, this process kicks out a titanic amount of energy. The Sun is heated to tens of thousands of degrees, kicking vast quantities of electromagnetic radiation out into the solar system in all directions. Objects orbiting the Sun such as the planets only receive a tiny, tiny fraction of this output, occupying just a tiny, tiny fraction of the space around the Sun. Planets are just motes of dust in the beam of a car's headlights, barely detectible from their star's perspective. Even so, the numbers of this energy transfer are quite literally astronomical.

Occupying a mere *fifty billionths of 1 per cent* of space visible to the Sun, the Earth still drinks in around 150 thousand trillion joules of sunlight every single second.

Throwing huge quantities of electromagnetic radiation out into space makes the Sun sound special, but, in fact, all objects in the universe do this. Physics tells us that every object constantly emits energy in the form of electromagnetic radiation, which we call *blackbody radiation*. The reason we don't see all objects glowing like the Sun, however, is that the quantity of blackbody radiation emitted depends on how hot the object is: the hotter the object, the more energy it emits per second. To be a little more specific, the quantity of energy emitted is proportional to the fourth power of an object's temperature, i.e., T^4, where the temperature is measured in kelvin.[2] Kelvin is the unit of absolute temperature, how hot an object is relative to absolute zero, and in practical terms is identical to the Celsius scale but with absolute zero as 0K rather than -273.15°C. This fourth power relationship means that even a modest increase in the temperature of an object results in it emitting a huge amount more blackbody radiation. The Sun, being so incredibly hot, thus kicks out colossally more energy per square metre than any everyday object on Earth.

The most important thing to understand about blackbody radiation is that, in addition to the *quantity* of energy emitted being dependent on the temperature of the object, so too is the *wavelength* of radiation emitted. You may have seen this for yourself if you've ever heated something in a fire.* Objects left in a hot fire for long enough will initially glow a dull, cherry red, then orange, and eventually, if the fire is really hot, a brilliant white. This is blackbody

---

* Perhaps this says more about me and my pyromaniacal years spent as a boy scout than intended.

radiation! We observe these changing colours because as an object gets hotter and hotter, it emits more and more short wavelength blackbody radiation, and less long wavelength radiation.* A tin can placed in a fire will initially glow red, emitting long wavelength red light, but after heating up some more it will start to emit shorter wavelength yellow light, mixing the two together to form orange light. Eventually, the can will be so hot that it even emits green and blue wavelengths of light, mixing together with the reds and yellows to form white light.

WAVELENGTH IN METRES (m)

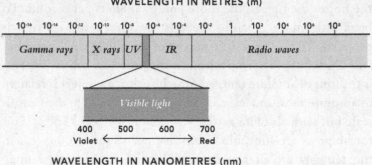

WAVELENGTH IN NANOMETRES (nm)

Figure 5: *The electromagnetic spectrum.*

To bring this back to our previous discussion, a relatively cool object like the Earth (having a temperature of a few hundred kelvin) doesn't emit much energy, and does so mostly at long wavelengths. More specifically, the Earth largely emits wavelengths in the infrared part of the spectrum, having wavelengths

---

* This is because short wavelengths of light carry more energy than long wavelengths of light, and so, to more efficiently transport heat away from an object, the hotter the object gets, the more it prefers to emit short wavelengths of light.

just a little longer than visible light. The Sun, however, being extremely large and extremely hot (thousands of kelvin at its surface), pumps out huge amounts of energy, largely at shorter wavelengths such as ultraviolet light. We will see that this difference in wavelengths between sunlight and 'earthlight' is of absolutely critical importance when talking about the atmosphere.

This may have all seemed quite abstract, perhaps a recurring feature of this chapter. But I think it was worth all the abstraction, for we now have all the conceptual tools we need to answer our original question: how and why does air temperature change?

Sunlight – mostly short wavelengths of light, fresh from the stellar forge – passes through the Earth's atmosphere and is absorbed by the planet underneath.* The Earth, drinking in this energy, heats up and then, like all objects in the universe, emits blackbody radiation of its own. As we just learned, however, because it is relatively cool, the Earth mostly emits longer, infrared wavelengths of light upwards, into the atmosphere. However, while the atmosphere is perfectly happy letting short wavelength sunlight pass through it, it's nowhere near as lenient when it comes to earthlight and its infrared radiation. In fact, our atmosphere appears like a brick wall to long wavelengths of light, and absorbs pretty much all of the energy the Earth emits as blackbody radiation. Having done this, the atmosphere heats up and then, of course, emits blackbody radiation of its own – half of it outwards, into space, and half of it inwards, towards the Earth.

This splitting of the energy into outward and inward halves is of

---

* The notable exception to this is of course the absorption of short wavelengths of light by ozone, as previously discussed when talking about the stratosphere.

critical importance to understanding global climate, but we will return to that idea much later on. For now, the key takeaway is that the atmosphere is not heated directly from *above* by the Sun, but instead heated from *below* by the Earth. To use a culinary analogy, the fluid of the atmosphere is heated from below like water in a pan on a stove, not from above like water in a pan under the grill.

This seemingly minor distinction has an important effect. When the water in the pan under the grill reaches boiling point, it merely steams at the surface, while on the stove it starts boiling, roiling with bubbles cascading to the surface. This is because under the grill our water is heated from above, meaning that the few millimetres of water closest to the surface are the warmest, with the few millimetres below being slightly cooler, and so on. When the warmest layer reaches boiling point, it simply evaporates, turning to steam and escaping, and exposing the next few millimetres of water to the air. The water possesses what we might call static stability.

The pan on the stove, however, flips this behaviour on its head. The water is now heated from below, with the bottom few millimetres of water being the warmest, then the few millimetres of water above being slightly cooler, and so on. When the water in the lowest, warmest layer reaches its boiling point, it turns into steam. Doing so, however, makes it less dense than its surroundings, and so it shoots up to the surface, where it pops and merges with the atmosphere as a bubble. Thanks to friction, these bubbles also drag along with them some adjacent water molecules, creating an overturning circulation – warm water is carried on updrafts from the bottom of the pan to the surface, and then, unable to also escape to the atmosphere, travels back down to the bottom in a return flow. If you've ever watched a pot boil* you

---

* If you plan on doing this, give yourself plenty of time.

will have seen this effect of circulation cells forming in boiling water, marked out by torrents of bubbles, constrained by the shape of the pan.

We see a very similar phenomenon in our atmosphere, minus the bubbles! And with our equation of state, we can very easily see why. Air at the surface is heated by the Earth below, but this heating isn't uniform – some areas will see higher air temperatures than others, such as air over land compared to air over water. Where air is warmer than its surroundings, it is also less dense. We know this because our equation of state says the pressure air experiences is proportional to its density multiplied by its temperature. As air is under pretty much the same pressure everywhere at the surface – pressure being nothing more than the weight of air above pressing down – this means that if the air temperature *increases*, air density must *decrease* in order to balance our equation of state. A bundle of less dense, warm air surrounded by more dense, cool air is then in much the same situation as a bubble of steam at the bottom of a pan of water on the stove – it rises!

We call this process *convection* (from the Latin 'to carry together'), and across the Earth we see vast overturning circulations driven by convection, just like the overturning circulations we see in a pan on the stove. Because the atmosphere covers the surface of a rotating sphere, rather than a stationary, cylindrical pan, however, these circulation patterns take on a very different shape. The largest of these patterns is centred over the equator, where air is heated more consistently than air at lower latitudes. This results in a planet-wide band of energetic convection, with air rising at low latitudes, then spreading out north and south. The return flow of this persistent, strong convection is so significant that we will spend much of the next chapter discussing it, and the impact it has had on not only the development of atmospheric science, but human society as a whole.

For now, however, I would like to take you back a few chapters to our daring aeronauts, Coxwell and Glaisher. I think it's time to finally explain why their flight was of such great interest to me, and what makes the stratosphere so special.

Glaisher has passed out. Coxwell is in the rigging of the balloon, desperately trying to free the line for the release valve. Next to the unconscious Glaisher, the hoar-frosted scientific instruments mounted on the wicker basket are a blizzard of activity. The barometer indicates that the air pressure is a third of what it was when the pair took off, and still dropping. The hygrometer reading has plummeted off the scale entirely. But among all the chaos and commotion, one instrument probably – and I say probably as Glaisher sadly was not cognisant enough to read it – was static. The thermometer indicated that despite ascending several hundred feet, the temperature in the balloon had not changed. We previously discussed how this temperature profile defines the stratosphere – air gets colder with altitude in the troposphere, and is initially constant in the stratosphere before warming with altitude.

Considering the discussion in this chapter, we can immediately see why air temperature in the troposphere decreases with altitude. With the atmosphere heated from below, much like our pot of water on the stove, the air closest to the ground is warmest. This air absorbs the blackbody radiation from the planet beneath, warms, and emits blackbody radiation of its own. In a way, the energy initially emitted by the ground is passed from each millimetre of the atmosphere to the next, starting at the surface and working all the way up to the edge of space. This process is somewhat leaky, however, with more and more energy lost at each handover as the air gets thinner and thinner (the 'lost' energy is radiated out to space). This results in the troposphere decreasing in temperature with altitude, as we would expect the whole atmosphere to. We of course know instead that the stratosphere does something completely different. We also only briefly touched on why this is the case – ozone.

Ozone – formally, O_3, or, less formally, three oxygen atoms stuck together – is the exception to the rule that sunlight passes through the Earth's atmosphere unscathed. Due to a quirk of how it is produced and destroyed (a balancing act of breaking apart O_2 molecules via radiation, and decreasing air density with altitude), ozone is concentrated in a thick layer in the Earth's stratosphere, and is scarcely found anywhere else naturally. The molecule is extremely good at intercepting ultraviolet light, so much so that certain wavelengths of ultraviolet light (UV-B, to be precise) are reduced in intensity by a factor of 350 million[3] between the top of the atmosphere and the bottom. By absorbing these particular wavelengths of light, ozone is a huge help to life on Earth, limiting the intensity of this cancer-inducing radiation at the surface. Were it not for the ozone layer, life on Earth would almost certainly look very different, limited by shorter lifespans. This is why the depletion of the ozone layer by human-produced chemicals such as chlorofluorocarbons (CFCs) in the twentieth century was so immensely concerning, though, at the time of writing, it appears that, thanks to international agreements, the ozone layer is on track to recover to its nineteenth-century state by the middle of the current century.[4]

This is good news for life on our planet, but also good news for the stratosphere, as ozone is what causes its unique temperature profile. As ozone molecules absorb ultraviolet light, they heat up, converting the energy of the radiation into thermal energy. Ozone is so abundant in the stratosphere that this heating is significant, countering the 'leaky' handover of energy from the surface, and causes enough warming to give the stratosphere its distinctive temperature profile.

However, this severely undersells why the stratosphere is so special. This bizarre property of increasing air temperature with altitude completely breaks one of the fundamental ways that we think the atmosphere works. To explain why this is, we need to start thinking about *air parcels*.

Imagine a balloon that is infinitesimally thin and weighs nothing, yet completely prevents heat transfer from the inside to the outside. Such a balloon thermally isolates a small volume of air, yet allows that volume to move around and deform totally unimpeded. Holding such a balloon in your hands, envelop a portion of air, perhaps the size of a dustbin, and seal it. Congratulations, you have created an air parcel! Obviously, it is a fantasy, but like so many concepts in science, it is a useful fantasy: air parcels allow us to keep track of a volume of air in a thought experiment, as we will now see.

You have an air parcel in your hands, and that, assuming you are not reading this on a long-distance flight,* means you are surrounded by tropospheric air. Let's assume for simplicity that this air around you is completely still and completely dry. Hoist your air parcel over your head, raising it up by maybe a metre, and let go.

By moving the air parcel vertically upwards we thrust it into an environment with a lower air pressure, which, of course, falls exponentially with height. The same weight of air above presses down on the air parcel and its surroundings, meaning that the air pressure is equal inside and outside the imaginary balloon. However, because we have thermally isolated the balloon, the temperature inside and outside the air parcel will not necessarily be equal. Recall that in the troposphere, air temperature decreases with altitude. Naively, this implies that our air parcel will be warmer than its new surroundings, originating from slightly

---

* Alternatively, if anyone is reading this right now in low Earth orbit or beyond, please, I'm begging you, let me know about it!

closer to the surface. In reality, air slightly cools as it ascends in a process called *adiabatic cooling*, something that happens regardless of us thermally isolating the air parcel. Depending on how quickly the surrounding air cools with altitude, our air parcel can be warmer than its surroundings, the same temperature as its surroundings, or cooler than them.

Cast your mind back to the equation of state for the atmosphere:

$$p = \rho \times R_s \times T$$

If our air parcel is at the same pressure and the same temperature as its new, elevated surroundings, we can conclude from this equation that it possesses the same density. Our air parcel simply stays where it is placed, with no force motivating it to rise any further or sink back down. To borrow a phrase from Douglas Adams, our balloon hangs in the air in exactly the way a brick doesn't. Things get a lot more interesting, however, if the air parcel is at a different temperature to its new surroundings. If it is warmer than its surroundings, yet still at the same pressure, the equation of state implies that it is *less* dense than its surroundings. Much like a bubble of less dense air in a swimming pool of very dense water, our air parcel is buoyant, and shoots away from us, upwards and upwards. We have convection! Convection occurs if the air temperature cools relatively quickly with altitude, or if the surface beneath is very warm. If the air temperature instead cools relatively slowly with altitude, our hoisted air parcel will then be cooler than its new surroundings, and so, as per the equation of state, also denser. Instead of shooting upwards, it gently returns to our hands, back to an environment that is the same temperature as its contents.

We can summarise all this by saying that the troposphere possesses *conditional stability*. Depending on the circumstances, air parcels in the troposphere can be unstable (resulting in convection) or stable (returning to their original location). Air here is unstable often enough, however, that convection is widespread, and the motion of the atmosphere three-dimensional.

Now, let's repeat the same experiment in the stratosphere. Climbing up an imaginary, 15-km-tall ladder, we trap a volume of air in our thermally isolating balloon and tie it off. Lifting it above our heads, we let go. The air parcel gently sinks back to where it started. Pack up the ladder, move to a different location, and repeat the experiment. Again, the displaced air parcel sinks back to its original location. You can repeat the experiment as many times as you like, but the result will always be the same. Because air temperature *increases* with altitude in the stratosphere, when an air parcel is displaced upwards, it will always be cooler than its new surroundings, and so denser, and thus will sink back down. We say that the stratosphere is *statically stable*. This means that vertical motion in the stratosphere is almost entirely prevented, with convection made impossible.

Therefore, had Coxwell and Glaisher been conscious (and not concerned with their rapidly approaching demise) at the zenith of their flight, they would have found themselves in an alien world. Here, a mere ten kilometres above the surface, air behaves differently. In the stratosphere, dynamics take place on flat surfaces – with no vertical motion, wind velocity is reduced to a two-dimensional flow at each level.* The inhibiting of upward motion also means that the stratosphere is almost entirely dry. Water can only enter from the troposphere below, but when moisture-laden air shoots up in a convective current,

* There is a grand, planet-wide overturning circulation in the stratosphere called the Brewer–Dobson circulation, but this is a very different beast to the massive, energetic convective circulations of the troposphere. See N. Butchart, 'The Brewer–Dobson Circulation', *Reviews of Geophysics* (2014), pp. 157–84.

it crashes into the statically stable stratosphere, unable to move any further (unless it is extremely energetic, which we do see sometimes). We can actually witness this very clearly when looking at cumulonimbus clouds – their distinctive anvil shape is caused by moisture-laden, upwardly mobile air hitting the tropopause and, unable to move any further upwards, spilling sideways, bringing the cloud with it. The flat tops of these clouds beautifully mark out the boundary between the stratosphere and the troposphere.

Compared to the dense, humid, three-dimensional troposphere, the stratosphere is a ghostly realm. I think I find it, and Coxwell and Glaisher's unintentional voyage into it, so fascinating because

Figure 6: *A cumulonimbus cloud – the flat top marks the boundary between the stratosphere and the troposphere, the tropopause.*

it is so close to us and yet so completely alien. If the stratosphere were at a horizontal distance rather than a vertical one, you could drive a car there in a matter of minutes, and on arriving you would be confronted by a realm of arid, flat motion. Yet for thousands and thousands of years, humans had no idea that their thin shell of an atmosphere was bounded above by this aeronautical desert. We were surrounded by an alien world for our entire history and yet had not the slightest clue that it existed. Our pair of pioneering balloonists stumbled into a scene of previously unimagined wonders, and, not quite prepared for the discovery, slipped back down through the clouds to safety.

We will return to the stratosphere and its unique circulation in a later chapter, but for now let's descend back to the troposphere. Here we will apply our new knowledge of atmospheric processes to what may be the most significant wind on the entire planet. Without it, the rich science of our atmosphere would simply not exist as we know it.

CHAPTER 5

TRADE

When Philip II of Macedon (382–336 BCE), father of Alexander the Great, went to war, he would set out in summer. This was not, as you might expect, to take advantage of fair weather for his battles, but because of a particular weather phenomenon in the eastern Mediterranean. From approximately May to August, a strong, dry wind persistently blows from the north to the south of the Aegean Sea. Philip went to war knowing that the powerful, rival city state of Athens would want to assist his enemies. This would be most effectively done by sea, and so he cleverly timed his operations such that any aid sent by Athens would be forced to sail into this strong wind, delaying their travel and effectively removing them from consideration.[1]

These winds, written about by Aristotle, became known as the Etesians, from the Ancient Greek for 'periodic [or annual] wind'. They are just one example of winds that form in a particular place at a particular time of year – other such examples in Europe include the Sirocco in the Mediterranean, the Mistral in France, and the Llevantades in Spain. Since antiquity, humans have identified patterns in atmospheric behaviour such as these, but for the longest time they lacked an accurate explanation for them. As we have already seen, originally supernatural explanations were invoked, while today we know that these winds are caused by an application of rudimentary physics to the immense, rotating fluid that is our atmosphere. All atmospheric marvels can be explained in much the same way, from hurricanes to cold fronts, heatwaves to aurora. The physics that underlie these events are in a way like the muscles and sinews of the great atmospheric giant, powering its footsteps that we experience as weather.

This being the case, why does the giant sometimes possess such a regular stride? Why does the atmosphere behave predictably in some cases, with winds like the Etesians flowing reliably every year, while in most cases the wind appears to blow at random? In order to understand how the giant can leave such deep, regular footprints, we must first understand the underlying mechanics. What motivates its muscles?

Previously we discussed how the flow of the atmosphere is ultimately caused by changes in air pressure. Other factors such as water vapour and terrain do come into consideration, but when looking at the atmosphere on a larger scale, by far the most important factor is air pressure. Air pressure is in turn forced by temperature; specifically, the temperature of the planet beneath it. And what forces the planet to be at a certain temperature, the ultimate cause of all winds on Earth? Of course, as we learned in the last chapter, our Sun.

But its energy is not distributed evenly.

Imagine that you have a square solar panel and lay it flat on the ground at the equator. The Sun, directly overhead, sees this panel head-on as a square, and illuminates this square with solar radiation. If, however, you were to take an identical solar panel, travel close to one of the Earth's poles and lay your panel flat on the ground, the outcome would be rather different. The Earth's surface curves away from the Sun as you move from the equator to the poles, and towards the night side. Or, in other, simpler words, the Earth is a sphere. Well, not quite, but we'll come back to this. As a result of this curvature, while the solar panel may still look square to you as you lie it on the snow, 150 million km away the Sun only sees a thin rectangular panel. While its width – running east to west – is unchanged, its height – running north to south – has been shortened by perspective. So instead of appearing square, the solar panel is foreshortened into an extremely thin rectangle. This means the solar panel occupies a smaller area of space seen by the Sun. And as our star pumps out an equal amount of energy into all space visible to it, this means that despite both panels being the same size, and nearly exactly the same distance from the Sun, the panel at the pole receives less energy than its counterpart at the equator.

This fundamental piece of geometry is why the equator is consistently warmer than the poles: it occupies more of the space visible to the Sun. Yet, as the year progresses, the Earth's axial tilt causes different areas of the planet to be exposed to the Sun to a greater or lesser degree. In the northern hemisphere's summer, the hemisphere is tilted towards the Sun such that it occupies more of the space seen by the Sun, and hence receives more energy. This causes the hemisphere to warm, bringing summer year after year. It also, however, brings certain predictable patterns in the atmosphere.

For example, in the northern hemisphere's summer, the arid deserts and steppes of western Asia scorch in the sun. Daytime temperatures can regularly soar over 40 °C, and long days, strong sunshine, and little moisture bring extreme heat to Iraq, Turkey, Syria, and Iran. As you may have guessed, this causes convection – air pressure has remained constant but air temperature increased, and so the air density decreases according to the equation of state.

Being less dense than the surrounding air, the warmer air is more buoyant, and so rises into the sky. This leaves behind a partial vacuum, and so the air pressure near the surface falls. As a result of this, a system of low pressure known as the Asiatic or Iranian low dominates western Asia in the summer months. Combined with the relatively high pressure in central Europe in summer (fuelled by a much moister climate), air is funnelled between the two systems in what is known as an atmospheric trough. Every year, this causes air to flow from the warm, mostly dry interior of eastern Europe towards the Mediterranean. This air flow is what we call the Etesian wind, blowing year after year, keeping Athens' fleet at harbour.

But, wait a minute. Why doesn't the air flow from the area of high pressure in Europe towards the area of low pressure in Asia? Naively, we would expect a west-to-east wind. Why do we end up with this southwards trough?

To find an answer to that question, we join a number of great scientists throughout history, and encounter another persistent wind in the atmosphere. This time, however, it didn't simply influence the fate of Philip II and his Macedonian armies – it shaped the history of the entire world.

Edmond Halley (1656–1742) has been remembered rather unfairly by history. His name is of course associated with the comet that returns to the night sky every seventy-six years or so, the only short-period comet that can be seen with the naked eye. Yet if you pressed most people to explain what else Halley achieved besides calculating something to do with a comet, they would struggle. A more extensive list of Halley's achievements is detailed in his splendid slate and gold memorial plaque in the South Cloister of Westminster Abbey. These include being the second Astronomer Royal, Fellow of the

Royal Society, and Savilian Professor of Geometry at Oxford University. Recognising its importance, and playing no small part in its creation, he also personally financed the publication of Isaac Newton's *Philosophiæ Naturalis Principia Mathematica*, quite possibly the most influential work of science ever written.* Also, unlike so many great men of science at the time, according to some accounts Halley was a barrel of laughs. The previous Astronomer Royal denounced Halley for drinking brandy and 'swearing like a sea captain', and one story even has him and Peter the Great, Tsar of Russia (1672–1725), taking it in turns to drunkenly push each other through hedges at a stately home in a wheelbarrow.[2] Yet the pithy line of his epitaph with the most relevance to our story reads simply: 'oceanographer, meteorologist, geophysicist'.

Like many great figures of the Enlightenment, Halley was only able to pursue a career in science thanks to the wealth of his family. While his father – a landowner and salt merchant – lost property in the Great Fire of London in 1666, there was still enough coin in the family coffers to tutor Edmond and send him to study at Oxford. For reasons lost to time, however, he did not stay there. Despite great academic success, including a publication while an undergraduate in the prestigious *Philosophical Transactions of the Royal Society*, in 1676, Halley abandoned his studies. Having studied at Oxford myself, I can

---

*   It should be noted that Halley paid for all the expenses of publishing the *Principia* because the Royal Society of London – of which he was clerk – was flat broke. Specifically, the Society spent all its money producing *De Historia Piscium* – a book by Francis Willughby (1635–72) on the history of fish. Perhaps not surprisingly, this sold spectacularly poorly, and so the Royal Society nearly robbed the world of the most important piece of science ever written due to falling hook, line, and sinker for a bad investment.

completely understand Halley's desire to get as far away from tutorials and lectures as humanly possible. I can honestly say, though, that I never once considered sailing to a remote volcanic island to do so. Yet this is what Halley did, with royal blessing. Using his father's connections, he obtained a letter from King Charles II (1630–85) that secured him passage on a British East India Company ship to Saint Helena, a speck of British territory in the middle of the south Atlantic.

Here he spent two years mapping the southern skies. Or at least, that was the plan. It turned out that the weather in the south Atlantic was rather cloudy, ruining night after night of observations. Halley found plenty else to do, however, and instead made notes on the atmospheric and oceanic conditions. On returning to England in 1678, he dedicated a planisphere (an articulated map of the sky) of the southern stars to Charles II and in return was granted the royal equivalent of a voucher to redeem one free degree from Oxford University, despite having not taken any examinations. It really is who you know, not what you know, that counts. Halley went on to pursue a dazzling range of research interests (including microscopy, archaeology, biology, engineering, astronomy, and mathematics) but came back to his meteorological observations several years later.

Halley's experience 'in an employment that obliged me to regard more than ordinary the weather'[3] led to a paper written in 1686, titled 'An Historical Account of the Trade Winds, and Monsoons, Observable in the Seas between and Near the Tropicks, with an Attempt to Assign the Phisical Cause of the Said Winds'. This was arguably the first academic paper ever written in the modern field of *climatology*, and was certainly a quantum leap forward in how the atmosphere was perceived by gentlemen scholars. Informed by a multitude of observers across the world, Halley reconstructed the wind patterns of the known world, and even attempted to display these wind patterns in a single map:

> To help the conception of the reader in a matter of so much difficulty, I believed it necessary to adjoyn a scheme, shewing at one view all the various tracts and courses of these winds; whereby 'tis possible the thing may be better understood, than by any verbal description whatsoever.

Figure 7: *Halley's map of global wind patterns from 1686.*

Halley assigns a particular direction and speed to the wind at a given location – instead of being random, the atmosphere is depicted as behaving predictably:

> I could think of no better way to design the course of the winds on the mapp, than by drawing rows of stroaks in the same line that a ship would move going alwaies [always] before it; the sharp end of each little stroak pointing out that part of the Horizon, from whence the wind continually comes.

With the benefit of modern satellite measurements, we can say that these wind patterns are broadly accurate, itself a remarkable achievement, though this was only half of what Halley attempted. Not content with simply describing the winds, he wanted to *explain* them. While Aristotle's incorrect argument that wind was the accumulation of dry, warm exhalation from the Earth still persisted into the seventeenth century, Halley was having none of it in his paper. 'Wind is properly defined to be the stream or current of the air, and where such current is perpetual or fixed in its course, 'tis necessary that it proceed from a permanent, unintermitting cause.' Halley was of the opinion that as the Sun warmed up a section of the atmosphere, the latter would expand and so become less dense. This section of atmosphere would then rise up, resulting in lower air pressure, and so air from other areas would rush in to replace it. This was how Halley explained one of the most important sections of the Earth's circulation, clearly identifiable on his chart: the *trade winds*.

The trade winds are permanent features of the Earth's tropical regions, where the atmosphere near the surface consistently flows from east to west. They can clearly be seen in Halley's map, blowing from the west coast of Africa to Mesoamerica, as well as from the Pacific into the Philippines. These reliable winds meant that certain parts of the globe became easy to access by seafarers, while rendering other regions inaccessible, shielding them from outside interference. One particularly important example of this is the ability of ships to sail from Europe to North America by the trade winds in the Atlantic Ocean. First recognised by Portuguese sailors in the early fifteenth century, these trade winds carried Christopher Columbus – and earlier travellers such as John Cabot – to the American continent. Resources brought back from the New World to Europe in the sixteenth and seventeenth centuries fundamentally redistributed global power.[4] Europe became the axis of the world. The wealth brought back to Europe directly and indirectly caused the scientific revolution, the invention of the barometer and thermometer, and the founding of what we would now call modern science. New World wealth was discovered using the trade winds, funded Halley's studies, and established the institutions that provided Halley with the data necessary to understand the winds. If the trade winds did not exist, it is extremely doubtful that atmospheric science would be as advanced as it is today.

Science is a process built on data. In the scientific method, one develops a hypothesis and then tests it against data. If the data supports the hypothesis, it can be considered correct until subsequent data repudiates it. As such, a scientist needs data to do their job – both to form hypotheses and to test them. Atmospheric science is unique in that, in order to understand the behaviour of the atmosphere on large scales, the necessary data must be drawn from locations separated by vast distances. For many phenomena to be understood, this data must be communicated from these great distances and collated into one dataset. For prediction of these phenomena to be possible, as we shall cover later in the book, that communication and collation needs to be near-instantaneous. While humans live all over the planet, as we have previously learned, they have only possessed the instruments necessary to measure relevant variables – such as air temperature and pressure – for a few centuries. Natural philosophers only shared data widely in fits and bursts before the invention of the printing press, and even subsequently have been limited by conflicts between nations. It is no coincidence then that the transformation of atmospheric science occurred at the same time as a transformation in global politics and economics, enabling global information flow as never before. The birth of *proto-globalisation*.[5]

Halley was conveyed to Saint Helena by the British East India Company. His groundbreaking 1686 paper was informed by observers all over the world as then known to Europeans. In fact, he says as much: 'It is not the work of one, nor of few, but of a multitude of observers, to bring together the experience requisite to compose a perfect and compleat history of these winds.' His theory of the trade winds was only made possible by collating correspondence from a huge number of dispersed observers in European overseas territories. As previously noted, the wealth transferred to Europe by the Columbian exchange and other colonial activities financed scientists and scientific institutions. This period is referred to as proto-globalisation, being the prelude to modern globalisation, the integration of people, companies, and governments across the world. The institutions of proto-globalisation such as colonial outposts and the East India Company made the development of atmospheric science possible through the most fundamental currency of science: data. Data was collected by ocean-going ships of transnational companies and collated in their headquarters. Data was collected in European colonies and collated by officials in organisations such as the Royal Society of London. The activities of proto-globalisation not only financed scientists but provided them with the data necessary to complete their studies. Atmospheric science arguably owes the greatest debt of all the sciences to this colonial structure, its development being entirely dependent on the extreme concentration of geographically dispersed information.

Science likes to think of itself as amoral, apolitical, objective. Individual scientists often are – they are curious individuals who wish to understand the world better, and will work with whatever data is available to them. However, the development of modern science was only made possible through the activities of European powers in the early modern period. This means that science owes a debt in data and money to – among other things – the triangular slave trade in the Atlantic, colonial invasions and suppressions, and international policy based on social Darwinism.

Modern data-driven science bears a heavy debt. This is not to say that individual scientists bear responsibility for these things. But consider the factors that enable Halley to make the following, enthusiastic plea in his 1686 paper:

> I shall take it for a very great kindness if any master of a ship, or other person, well informed of the nature of the winds, in any of the aforementioned parts of the world, shall please to communicate their observations thereupon; that so what I have here collected may be either confirmed or amended, or by the addition of some material circumstances enlarged.

Had history played out differently, perhaps atmospheric science would have arisen through a more humane, equitable route. But history did play out this way. We must acknowledge the colonial, often brutal way in which early modern science acquired its data.

Informed by this data, Halley was nearly correct in his explanation of the trade winds. He argued that as the Sun passed from east to west in the sky, warm air heated by the sunlight would rush in from behind the path of the Sun, forcing an east-to-west circulation in the atmosphere. He was nearly there, but not quite. Being a brilliant scientist, he himself realised this. If the explanation of the trade winds was there in the data, Halley, in his fumbling through, had brushed his fingertips against the correct explanation. He implored others to take on the challenge, rummage through the data, and finish his work. The man who would plunge both hands into the lucky dip and emerge grasping the solution fifty years later was, of all things, a rather bored lawyer.

George Hadley (1685–1768) studied law at Oxford University and was called to the bar in London at the start of the eighteenth century. Yet his main passion was the nascent science of meteorology, and for seven years he was in charge of interpreting meteorological reports sent to the Royal Society of London from around the world. So, much like Halley, he had ready access to a widely dispersed network of observations and, much like Halley again, he used this data to develop his own theory of the trade winds. In 1735, shortly after being elected a Fellow of the Royal Society, he published a slim paper on the subject, 'Concerning the Cause of the General Trade Winds'.[6] His improvements to Halley's theory were twofold. First, he correctly dismissed the idea that warm air rushed in from behind the path of the Sun. After all, if this were the case then the whole planet would experience persistent easterly winds. Instead, Hadley asserted that when air was heated in the tropics and ascended, it was replaced by air not from the east, but from both north and south of the equator, where the heating by the Sun was marginally less. This air rushed towards the equator, where it converged. Hadley's second improvement to Halley's theory, and it was a particularly clever idea, was that the rotation of the Earth would affect these converging air masses.

We will all have experienced those hot days on which the atmosphere just feels . . . heavy. Thick air surrounds us, unstirred by even the slightest breath of wind. Relative to us on the Earth's surface, the local atmosphere is stationary. However, that's just our perspective. A cosmic observer, hanging above the planet in a geostationary orbit, would see the atmosphere hurtling around at hundreds of miles per hour. This is, of course, due to the Earth's rotation. Completing one revolution every twenty-four hours, the Earth's surface travels through space at some 1,100 miles per hour at the equator, decreasing to 0 miles per hour at the poles, where the surface intersects with the axis of rotation. A cosmic observer would note that the atmosphere over the equator – dragged along by the Earth's surface – moves faster than the atmosphere further north and further south. Hadley reasoned that when air rushes into the equatorial region to replace the newly upwardly mobile air, it would move more slowly than the air at the equator. Specifically, as the Earth is rotating from west to east, the air moves more slowly *eastward*. An observer on the surface would therefore perceive this decrease in eastward flow as a *westerly* wind.[7]

If this is a little difficult to picture, think of it like this. Imagine that you are driving along a two-lane highway, with a low-speed and a fast-speed lane. Driving in the first lane you see cars in the second lane travelling faster than you. If you change from the first lane into the second while maintaining a constant speed, you are now travelling slower than the other vehicles in your lane. A stationary observer on the side of the road would see all vehicles in the second lane travelling in the same direction at high speed. But to the driver of the vehicle in front of you, because you are travelling at a lower speed in the same direction, you are actually moving away from them in their rear-view mirror. *Relative to their car*, you are travelling in the opposite direction. This is exactly what happens when an air parcel 'changes lanes' and moves towards the equator – because the air parcel is travelling at a lower eastward speed compared to the air flow at its new latitude, it is perceived by observers at that latitude to move in the opposite direction: westerly.

Hadley's argument here is pretty remarkable! Using his experience collating meteorological data, combined with excellent physical intuition, he had solved the mystery of the trade winds. In his honour, we call the overturning circulation that reaches from the equator to around 30° north and south, causing the trade winds with their return flow, the Hadley cell.

Well, Hadley had almost solved the mystery. There was one extra factor to consider.

Imagine that you are again high above one of the Earth's poles and looking straight down at the ground. If you were to ascend above the surface and speed up time (and also neglect friction with the surrounding air, but as a physicist this goes almost without saying), you would see the Earth rotating directly beneath you. At the north pole, the planet would be rotating anticlockwise, while at the south pole the planet would be rotating clockwise. Now, travel to the equator and repeat the same experiment. Ascending into the tropical air and again speeding up time, you would see no rotation directly underneath you. Yes, the planet would *move* underneath you, but it would do so in a linear fashion – from west to east. There would be no rotation or twisting of the landscape beneath you. In other words, as you travel away from a rotating planet's equator, said rotation gets more and more important. Perhaps the best way to think about this is in terms of *angular momentum*.

In school physics lessons we learn about momentum, the product of an object's mass and its velocity. This is in fact only one form of momentum: *linear momentum*. To use highly scientific language, linear momentum is an object's 'movingness'. A light object has less movingness than a heavy object travelling at the same speed. Equally, an object moving faster than another otherwise identical object has more movingness. To be more specific, when talking about moving an object, we are talking about moving the object in a straight line. As soon as you start to curve the trajectory of an object, you introduce a new kind of momentum with slightly different properties. This is the *angular momentum*. Extending our extremely technical language, angular momentum can be thought of as an object's 'turning-ness'. An object rotating about an axis will have more turning-ness if it rotates faster, making more revolutions per second. Equally, an object will have more turningness if it rotates at a greater distance from its axis. And, just as with linear momentum, an object rotating about the same axis at the same rate as another object will possess more turningness if it is more massive.

The two kinds of momentum are used so widely in physics because of a useful property they share: they are conserved. This simply means that an object – or a collection of objects – will always have the same total momentum unless acted on by a force, or a torque in the case of angular momentum. So, if an object with a certain amount of angular momentum were to start decreasing its distance from the axis of rotation, in order to conserve its angular momentum it must start orbiting faster. This is something that you can test yourself if you own a spinning office chair – set yourself spinning with your arms extended and a heavy book in your hands. If you then draw the book tight to your body, you will notice an immediate increase in your rotational speed – in order to conserve your angular momentum, you will spin faster! Equally, if a rotating object were to drift further from its axis of rotation, in order to conserve its angular momentum it would need to rotate more slowly. This is the rotational equivalent of the conservation of momentum that is taught in school physics classes when examining collisions. We can predict what will happen after an event, whether that's a change in an object's orbit or a collision between idealised objects on a frictionless plane, using the same principle of conservation of momentum.[*]

[*] These conservation properties are a consequence of one of the most beautiful theorems in mathematics, derived by Emmy Noether (1882–1935). To put it simply, linear momentum is conserved because physics is the same no matter where you are in the universe, while angular momentum is conserved because physics is the same no matter which direction you look in. This is an incredibly powerful result that proves very useful in constructing theories in physics. See, e.g., D. Neuenschwander, *Emmy Noether's Wonderful Theorem*, Baltimore, MD: Johns Hopkins University Press, 2017.

Let's revisit Hadley's air converging at the equator. In fact, recall that in Chapter 4 we discussed air parcels – the theoretical isolated bits of air useful in abstract experiments in the atmosphere. Let's dust them off and apply them here to get a sense of what's going on. As Hadley correctly worked out, air parcels are travelling from a region with a smaller distance to the axis of rotation, and are therefore moving slower – even at a standstill relative to the Earth's surface – than the air at the equator. This conservation of linear velocity produces an east-to-west wind. What Hadley neglected in his calculations, however, was that the moving air parcels also need to preserve their *angular* momentum. As the air parcels are increasing their radius of rotation by moving equatorward – there is a greater distance to the Earth's axis of rotation at the equator than at points further north or south – to conserve their angular momentum they must slow down. This momentum-conserving deceleration results in trade winds that are slightly weaker than Hadley calculated. This effect was not known in the early eighteenth century, and would only be mathematically expressed some hundred years later by Gaspard-Gustave Coriolis (1792–1843).

Son of an army officer, Coriolis had a glittering academic career, excelling in mathematics from a young age and eventually becoming a professor of mechanics. He formed part of a truly remarkable collection of French mathematicians all working in Paris at the same time: Cauchy, Navier, Galois, Laplace, and Fourier, to name just a few. He was instrumental in the development of modern mechanics, and introduced the terms 'work' and 'kinetic energy' to physics,[8] but is of course better known for introducing the idea of a 'Coriolis force' or 'Coriolis acceleration'.

As we just saw, this force (or, equivalently, acceleration – thanks Newton) is a result of an object attempting to conserve its angular momentum, which will alter its orbit around an axis of rotation.

For example, an air parcel that moves northward from the equator is deflected to the right, i.e., eastward. This is, however, in no way limited just to motion on planets' surfaces. The effect occurs whenever an object moves relative to any rotating reference frame. In fact, when Coriolis published 'On the Equations of Relative Movement of Systems of Bodies' in 1835 he did not even reference the atmosphere or the rotation of the Earth once – instead examining the transfer of energy in general rotating systems, using waterwheels as an example.[9]

This deflection of north–south motion by the Coriolis acceleration is well known, and absolutely essential to understanding how our atmosphere behaves. But, as an aside, there is another, far less known acceleration that affects east–west motion. This doesn't greatly affect atmospheric dynamics, but is a rather interesting parallel to the famous Coriolis deflection, and, strictly speaking, is also an easy way to lose weight fast!

As we just learned, air moving away from the equator is deflected to the east, e.g., a parcel of air trying to move north-wards in the northern hemisphere ends up moving north-east. This happens because the distance between a parcel of air and the Earth's axis of rotation decreases as the air parcel moves away from the equator, and so to preserve the angular momentum of the air parcel it must increase its speed in the direction of rotation. However, instead of moving an object in the north–south direction, let's now consider moving one in the east–west direction. Doing so we are not changing the object's effective radius of rotation, as we are not changing its latitude. We are, however, changing the velocity that it rotates about this axis with. If our object was initially stationary with respect to the Earth's surface, relative to the Earth's axis it was rotating with the velocity of the Earth's surface. This means its velocity was large if it was at the equator, small if it was near the poles. If we set it in motion in a west-east direction, the object is now rotating about the axis at this initial speed plus the object's speed relative to the Earth's surface. For example, if we were at the equator then initially the object – stationary with respect to the ground – had a speed of around 1,100 miles per hour relative to the axis. If we set it moving at 30 miles per hour – again, relative to the Earth's surface – in a west-to-east direction then it would have a speed of 1,130 miles per hour relative to the Earth's axis of rotation. Equally, if it were to move at 30 miles per hour in an east-to-west direction then it would have a speed of 1,070 miles per hour with respect to the axis.

By changing the object's speed relative to the Earth's axis of rotation, we are introducing an additional acceleration acting on the object – a centrifugal acceleration. This is an acceleration that seeks to change the object's radius of rotation about the Earth's axis – if the object increases its speed, this acceleration acts to push the object to a further orbit, while if the object decreases its speed then the acceleration acts to reduce the object's orbit. As this acceleration acts radially away from the Earth, in the exact opposite direction to the acceleration due to gravity, which acts radially inwards, we can say that the object's east–west velocity changes the weight of the object!* If the object starts moving east then the resulting centrifugal acceleration acts to partially counter the pull of gravity. A smaller acceleration pulling the object down is then felt as reduced gravity. So, when moving east, you would technically weigh slightly less – though not by much, it should be said: if you're travelling, say, along a British motorway at 60 miles per hour due east – perhaps on the M4 from Bristol to

---

* The weight of an object is, after all, its mass multiplied by the gravitational acceleration it experiences.

London – you would weigh only about 0.03 per cent less than if you were stationary.

This is known as the Eötvös effect, after Hungarian physicist Baron Roland von Eötvös (1848–1919). If you wanted to take this effect to the extreme, you could calculate the velocity necessary to produce a centrifugal acceleration equal to the acceleration due to gravity, i.e., to render you weightless. Doing so, you find an answer of around 17,400 miles per hour – which just so happens to be the speed of objects in low Earth orbit, such as the International Space Station! In other words, an object reaches orbit when it is travelling east fast enough for the Eötvös effect to totally counter the effect of gravity.

In a just world, the name of Eötvös should be just as well known as that of Coriolis, but sadly he (and the incredibly pleasing mathematical derivation of his eponymous acceleration)* are known to just a handful of Earth scientists. I'd like to encourage you to go out and spread his name! And tell people that if they want to lose weight, the easiest thing to do is to just go running – east. Really, really fast.

* I made a video about this! Search YouTube for 'why you weigh more when travelling east'

CHAPTER 6

DISTANCE

Some countries have *climate*, while other countries have *weather*.

This is a phrase I was introduced to while writing this book, and in a pithy way, it's incredibly accurate. Compare, for example, the British Isles and California. The weather in the UK and Ireland changes frequently, day by day, often hour by hour. Packing for a day out can require a variety of outfit changes to match the varying rainfall and temperature. Any given set of weather conditions rarely lasts longer than a couple of days – the British Isles definitely possess weather. Contrast this with the conditions in Californian cities like Los Angeles and San Diego, where each day is mostly like the previous one. Cloud cover may change, temperatures may fluctuate, fog may even make an appearance. But, compared to the weather in the British Isles, California possesses a climate.

Of course, all countries have *both* weather and a climate. Climate is simply the long-term average of weather, which is itself the day-to-day variations in atmospheric conditions: temperature, humidity, cloud cover, and so on. Edinburgh, for example, has what we call a temperate maritime climate, characterised by proximity to the Atlantic Ocean, while Los Angeles has a Mediterranean climate, mostly warm with seasonal changes in rainfall.[1] But, in living in Edinburgh and LA, you experience more acutely the short-term changes in the former, and the long-term changes in the latter.

I believe it's no coincidence that the majority of atmospheric science as we know it today originated in locations that have weather, not climate. From Torricelli in Italy to Ferrel in Massachusetts, both of whom we've already met, to FitzRoy in England and Bjerknes in Norway, who we have not, a common thread in the origin stories of atmospheric scientists has been meteorological unpredictability. When the world outside your window is pretty much the same from one day to the next, there is little incentive to wonder quite how it works. A world that has mood swings every couple of hours, on the other hand, is a mystery worth solving!

In particular, Western Europe is one of the most fascinating regions in the world from a meteorological perspective. The weather here is fiendishly difficult to predict, perhaps most of all in the British Isles. We've already discussed how the concentration of wealth enabled by the colonisation of the Americas, coupled with the institutions of global trade, enabled the birth of atmospheric science in Europe. But I would contend that this extreme variability in local weather is perhaps why the science flourished after its birth. Finally given the right tools, scientists had a mystery to solve, and they went about it with vigour.

But why is it that Western Europe has weather, not climate? Why does the world outside the windows of London and Oslo change with such great frequency?

The answer is the jet stream.

In simple terms, a jet stream is a narrow ribbon of fast-moving air, travelling from west to east around the globe. More practically, if you live in the mid-latitudes, the jet stream determines much of the weather that you experience. It's of incredible importance to weather prediction and behaviour of the atmosphere overall.

Earth has two jet streams in each hemisphere, one around 30° from the equator, and another about 60° from the equator. The first of these – the subtropical jet – occurs as the Hadley circulation descends and is deflected by the Coriolis effect to the east. This results in a strong, consistent westerly wind at around 10 km in altitude. The other jet stream in each hemisphere is stronger, and of greater relevance to our story. This is the polar jet, sometimes called the mid-latitude jet. It is an example of an 'eddy-driven jet', which is to say that it arises as a result of complicated interactions between alternating areas of high and low pressure causing wind to converge in the mid-latitudes. This jet snakes around the world, meandering in great loops and influencing a huge amount of the weather felt underneath. Sometimes in the northern hemisphere it will travel far enough south to connect with the subtropical jet, leaving Earth with just one jet north of the equator. If the planet were slightly larger, the two jets would be far more distinct – in fact, assuming a constant rate of rotation, the larger a planet gets, the more jets appear in its atmosphere. At the extreme end of the scale, Jupiter has no fewer than seven jets in each hemisphere!

Despite being of great importance to weather in the mid-latitudes, and not being too far from the surface, the jet stream is a relatively recent discovery – first identified in the mid-twentieth century. Historically, the discovery of the jet stream is ascribed to the Department of Meteorology, University of Chicago, which published a landmark paper in 1947, unusually attributed not to an individual scientist or a list of scientists but simply to 'Staff Members of the Department of Meteorology'.[2] If you were to assemble a fantasy team of mid-twentieth-century meteorologists, your roster would look remarkably like the Chicago Department (1947–48), headed by the legendary Carl Rossby (1898–1957).[3] This research was informed by the experiences of Allied bomber crews over both Europe and the Pacific in the Second World War, at times flying much faster or much slower than expected, due to the hitherto unknown fast air speeds at high altitude.

However, this attribution is wrong for two key reasons. First, the term 'jet stream' had already been introduced to meteorology in 1939 by German scientist Heinrich Seilkopf (1895–1968). Second, more importantly, the first paper on the mid-latitude jet was published more than a decade before the Chicago one.[4] This paper was remarkable, and tragic, for a number of reasons.

Wasaburo Ooishi (1874–1950) was perhaps too talented for his own good. In 1920 he was appointed as the first director of the Mount Tsukuba observatory in Ibaraki prefecture, Japan, and here he embarked on a programme of upper-atmosphere observations, inspired by the weather balloons of Richard Assmann.[5] By releasing large hydrogen-filled balloons, local wind speed and direction could be determined by tracking their paths with telescopes that could measure vertical and horizontal angles. Over many years and more than a thousand balloon launches, Ooishi had discovered something: the winds at around 10 km in altitude over Japan were extremely strong, sometimes as fast as 70 metres per second (250 kph or 155 mph). Ooishi's balloons were consistently swept due east into the Pacific Ocean at incredible speeds. Ooishi recognised that this was an important discovery and so should be shared as widely as possible. He excitedly wrote a scientific paper with the explicit purpose of maximising the impact of his ground-breaking research. Knowing that few scientists were familiar with the Japanese language, Ooishi didn't write in his first language but instead in his second.

Unfortunately for Ooishi, and the world, this wasn't the academic lingua franca of English but the constructed language, Esperanto.

Created in 1887, Esperanto is an international language designed to be easy to learn, with the noble intention of fostering world peace. However, even with a modern revival of interest, the language still has fewer than two million speakers, with considerably fewer in the 1920s. One of them was Ooishi, who, as well as being a high-flying meteorologist, was a passionate Esperantist, and would in fact later serve as the president of the Japanese Esperanto Institute. Doubtless sensing an opportunity to combine his two passions, he presumably thought that publishing such an important discovery in the language would bring it widespread attention. As it turned out, the paper was entirely ignored in the wider world. No fewer than eighteen further publications on the subject, all written by Ooishi, all written in Esperanto, were ignored as well.

His work did not go unnoticed in Japan, however. During the Second World War the Japanese military exploited their knowledge of the jet stream to launch 'Fu-Go' balloons – hydrogen balloons equipped with anti-personnel or incendiary bombs, released into the jet stream in Japan and then timed to release their payload over the United States. The intention was to instil terror in American civilian populations and wreak havoc by causing wildfires. As a weapon, the balloons were ahead of their time, the first with an intercontinental range. They were, however, rather ineffective. Only around 3 per cent of the balloons are thought to have made it to the North American continent (though this equated still to about three hundred potential bombing incidents), as data at the time slightly overestimated the speed of the jet across the entire Pacific, meaning most balloons harmlessly released their payload prematurely into the ocean.[6] As Tim Woollings notes in *Jet Stream: A Journey Through Our Changing*

Climate, there were only two notable strikes by these balloons.[7] The first was at a Sunday school picnic in Oregon, where six people were killed. These six were in fact the only casualties of the war on the US mainland, and were at the time the longest-ranged casualties in the history of warfare. The other balloon strike damaged the power supply of the Hanford nuclear weapons plant in the state of Washington. Within a year, this factory produced the plutonium that formed the nuclear payload of the American counter-attack on Nagasaki.

The Fu-Go balloons are now regarded as a historical curiosity, having had very little impact on the course of the war. Had their payloads been timed to release a little later, informed by more extensive jet stream data, perhaps the conflict in the Pacific might have gone rather differently. Equally, however, if Ooishi had published his results in Japanese or English, and Allied forces had been aware of the tremendous wind speeds in the upper troposphere, then bombing campaigns in both Pacific and European theatres would have proceeded very differently. As was realised outside of Japan after the Second World War, the jet stream has a massively important role in the weather of the mid-latitudes. In fact, when viewed on large scales, the behaviour of the troposphere here is principally the behaviour of the jet stream.

The jet influences European weather, and indeed all weather in the mid-latitudes, by influencing *weather systems*. These are neither small localised weather events such as a rainy downpour or fog, nor a consistent feature of the atmosphere such as the trade winds. Weather systems are the ever-changing moods of the atmospheric giant, streaming and rolling over the terrain of the Earth underneath, sometimes treading lightly and at other times crushing down with devastating force. They can be a few kilometres or a few thousand kilometres in scale, be hot or cold, dry or wet, but all obey the same physics and the same few equations.

The weather systems that affect Western Europe are primarily determined by the configuration of the mid-latitude jet stream.* The position and waviness of the jet can block certain air masses from entering the region, or draw in heat and moisture from elsewhere. The jet stream does not *create* weather per se, but instead acts as a master of ceremonies – dictating what weather conditions are experienced, outsourcing the creation of weather to other regions. For example, when the jet weakens and meanders far north of Europe, an area of high pressure dominates the area. This causes prolonged periods of the same weather, typically dry and warm in summer, due to other air masses, colder and bearing moisture, being unable to enter the region. For this reason, we call these events 'blocking highs'.

Generally, the jet stream 'imports' weather from the west, as both the jet itself and the meanders in its path progress from west to east. This explains why Western Europe, and the British Isles in particular, are largely temperate and wet – prevailing conditions come in from the west, Atlantic air that is wet and temperate year-round. This is the long-term average, however, and the high variability in European weather comes from the ever-changing configuration of the jet stream. If the jet waves further south, cold Arctic air plunges into Europe. If the jet plunges southwards in the Atlantic, the return flow can entrain warm African air over Europe (sometimes even bringing Saharan sand with it).[8]

* This is of course a complex issue, and I am oversimplifying by ascribing most of European weather to the jet stream! Quite apart from other large-scale factors such as teleconnections, which we will come to in a few pages, this discussion also doesn't factor in such local factors as orography, land use, and so forth. These will be discussed in greater detail in the next chapter.

Our journey zooming out from *in situ* weather is not complete, however. We have progressed from initially considering local factors to regional factors affecting weather, but there is one remaining scale: global. What, after all, does the jet stream respond to changes in? If events hundreds of kilometres away can influence weather, can events thousands of kilometres away? To answer that question, we need to go to the other side of the world.

Monsoon (from the Arabic *mawsim*, meaning 'season') refers to a regular pattern of seasonal winds and rainfall that takes place in the tropical Indian Ocean.[9] The monsoon is one of the most important atmospheric cycles on the planet, crucial to the livelihoods of billions. 'What the four seasons of the year mean to the European, the one season of the monsoon means to the Indian', Khushwant Singh writes in *I Shall Not Hear the Nightingale*. 'It is preceded by desolation; it brings with it hopes of spring; it has the fullness of summer and the fulfilment of autumn all in one.'[10]

The monsoon occurs due to the north–south asymmetry of the Indian Ocean: north of the equator lies the Indian subcontinent, while south of the equator there's nothing but water until you reach Antarctica. In the northern hemisphere's summer months, the land heats up faster than the ocean, and so a temperature gradient forms from north to south. This temperature gradient then drives an atmospheric circulation that carries warm, moisture-laden air over the ocean, first westward and then northward. This transported air then dumps its moisture over the Indian subcontinent, providing around 80 per cent of India's yearly rainfall in just a few months.[11] Monsoon rains initially fall unceasingly for days; then, as the season progresses, they fall for a few hours most days before tapering off entirely. In winter, the circulation reverses and the subcontinent is almost entirely without rain. The arrival of the monsoon rain, then, is a momentous occasion, the sudden shift from hydrological famine to feast transforming a parched landscape into a luscious one. Farmers depend on it, and anxiously await the monsoon's arrival. The timing is crucial, as the land can only go for so long without water. Should the monsoon be late, even by a few weeks, or bring less rain than normal, crops – and people – die.

For this reason, the monsoon has been studied for millennia. Epic poems and ancient philosophical texts from India make references to the seasonal rains, and much as with meteorology in the ancient Middle East, there was an interplay between folklore, natural philosophy, and religion. This would be replaced by a more statistical, Westernised approach, however, with the occupation of the subcontinent by the British.

Over the second half of the eighteenth century, the British Empire began to force its influence on present-day India, almost entirely through the East India Company, assisted by state armed forces.[12] By the middle of the nineteenth century, Britain's control over the subcontinent was absolute. The vast colony was the crown jewel in a global empire, its economy geared towards making money for Britain. To drastically oversimplify a complicated colonial history, the British colonial office in India was concerned with profit, not people. An estimated $44.6tn was extracted over the entire colonial period.[13] The local population suffered greatly during this time, never more so than in great famines, such as in 1876, 1896, and 1902. In the time that trillions of dollars of wealth were squeezed from the colony, an estimated 60 million deaths occurred due to famine, many directly the result of British economic policy.[14]

Members of the British colonial government tasked with improving understanding of the monsoon – after all, how could the colonial government earn money if the monsoon unexpectedly failed? – were able to identify the root cause of the seasonal rains as being the equatorial current driven by the temperature difference in the ocean. A horrific drought in 1899, unpredicted by colonial meteorologists and causing millions of deaths, however, showed that there was something else at play. An additional external factor was causing the monsoon to fail. Understanding what this could be was a matter of life and death for millions.

It would turn out that this external factor was on the other side of the world. And, not content with influencing the fate of a subcontinent, it would prove to be the most significant climatic variation on Earth. It may occur a world away but its mysterious influence was first understood in India. It's name? El Niño.

El Niño was first written about by Spanish colonisers in Peru some four centuries ago, though there can be little doubt that the indigenous population – who never developed a written alphabet – were already aware of the phenomenon.[15] The colonisers noticed that certain years were characterised by heavy rainfall, warm seas, and blooming vegetation, while others were barren and parched. The years of bounty were made possible because of a warm current appearing along the Pacific coast from the north, initially around Christmas-time, bringing warmth and moisture to the local environment. Because of this timing, local fishermen referred to the current as the (Christ) child, or, in Spanish, El Niño. Some years saw a strong effect, with a strong Pacific current and lots of extra rainfall; others were more subdued, with less of a warming in the ocean and less rainfall on land. Some years even brought the exact opposite, with a cold current in the Pacific and less rainfall on land. With an admirable level of conceptual symmetry, the colonisers referred to the cold current as *La Niña*, or the young girl.

While in the Spanish colonial period the appearance of the El Niño current, and associated wet weather, was considered a blessing, as population densities grew, strong El Niño events came to be feared. The Christ child now brings rainfall that sweeps away bridges and sets off mudslides that can kill hundreds. But these are only the local effects. A strong El Niño is felt quite literally the world over, and perhaps nowhere more keenly than in the Indian subcontinent.

The man who first unpicked its mystery was colonial officer Gilbert Walker (1868–1958). Born of privilege, Walker studied mathematics at the University of Cambridge, and possessed wide-ranging interests, from statistics to watercolour painting to playing the flute.[*] His fascination with flight, and boomerang flight in particular, earned him the nickname 'Boomerang Walker' while at Cambridge. Walker was appointed to foreign service in India in 1903, and there turned his attention to the monsoon. Unusually for this book, he wasn't trained as a meteorologist, but instead trusted the use of statistics to establish relationships between the monsoon and other phenomena. Over fifteen years of work, meticulously analysing the weather data available to him and calculating by hand the correlations[†] between different meteorological conditions, Walker eventually identified what he called 'strategic points of world weather'.[16] One of these strategic points was in the North Atlantic, what we now call the North Atlantic Oscillation or NAO (more on that in the next chapter). Another, hitherto unknown, Walker christened the 'Southern Oscillation'.

In simple terms, the Southern Oscillation is a sloshing of air mass between the Pacific and Indian Oceans. For convenience,

---

[*]   He actually made some design changes to the modern flute. See P. Sheppard, 'Obituary of Sir Gilbert Walker, CSI, FRS', *Quarterly Journal of the Royal Meteorological Society*, vol. 83, no. 364 (1959).

[†]   In simple terms, a correlation is an 'if this, then that, relationship' – two variables are *correlated* if, when one is observed to increase in value, the other increases in value also.

however, we typically measure it as a see-saw over the eastern and western parts of the Pacific Ocean (in Tahiti and Darwin, Australia, respectively). By a see-saw or a sloshing, I mean that when the pressure is relatively high over one area, the pressure will be relatively low over the other. As we know from previous chapters, such a pressure imbalance inevitably gives rise to an atmospheric circulation, and in this case we call the resulting air current the *Walker circulation*. This circulation of course transports air across the Pacific Ocean, and we can describe which way the circulation is flowing (east to west or west to east) via a number, the Southern Oscillation Index, which also tells us how strong the circulation is (the larger the number, positive or negative, the more strongly the circulation flows).

Through his meticulous statistics, Walker realised that the value of this Southern Oscillation Index influenced the Indian monsoon. In particular, when the pressure was anomalously low over the eastern Pacific, the Indian monsoon was likely to fail. Walker had no physical explanation for this, and lacked the relevant equations to describe what his statistics were telling him. The Southern Oscillation Index varies reasonably slowly, however, and so this knowledge was still valuable in anticipating drought, and hence famine, conditions in India. Knowing that the average pressure in the eastern Pacific was lower than normal gave authorities precious time to prepare for potential disaster.

Yet how did the Southern Oscillation connect to the ocean current off South America? It would take another half-century for equations to be derived that allowed the cause of this coupling to be understood, passing the baton of connecting El Niño to the monsoon to the Norwegian meteorologist Jacob Bjerknes (1897–1975). Bjerknes realised that El Niño and the Southern Oscillation were interlinked in what would become known as the *El Niño Southern Oscillation*, or *ENSO*.[17] This is a wonderful, beautifully complex system consisting of both atmospheric and oceanic components.[18] ENSO is a duet between the two largest fluid dynamical systems on Earth, spanning the largest ocean basin on our planet, with impacts that stretch across the globe.

It starts with the Walker circulation. Air flows en masse over the equatorial Pacific from east to west (recall the trade winds), moderated by variations in atmospheric pressure over the Pacific. As this air flows east to west across the Pacific, it causes an ocean current to develop that similarly transports water from the eastern part of the ocean to the west. Along the way, this water is heated by the Sun, such that when it arrives in the western Pacific it can be significantly warmer – around 8 to 10 °C – than when it left the eastern Pacific. We therefore have two currents, two conveyor belts of water and air, running in parallel. Because of the gradually heated conveyor belt of water, the western Pacific is much warmer than the eastern Pacific. This is compounded by the fact that as water is drawn away from the eastern ocean, it is replaced by water drawn up from the depths of the east Pacific. This water, coming from the deep, is much, much colder. It is also, however, stuffed with nutrients, and so perfect for sustaining a large population of fish. This is why fishing grounds are so rich off the west coast of South America, fuelling the rise of multiple civilisations.

The Walker circulation then indirectly transports heat across the Pacific, removing it from the east and bestowing it on western waters. As well as this, however, it also transports moisture. As the water beneath the air current is warmed, it evaporates and increases the humidity of the air. Eventually, the air contains too much moisture and it is dramatically deposited in extensive rainfall in the western Pacific. This joint transport of moisture and heat then spills over into the next ocean basin, rolling over Indonesia, directly and indirectly affecting conditions in the Indian Ocean. In particular, the Walker circulation acts to alter the temperature of the ocean. This changes the amount of evaporation that takes place and where any evaporated water is transported – normally west and then north, over India, to form the monsoon. In other words, the Walker circulation does not *cause* the monsoon, but *enables* it, and changes in the circulation can have a knock-on effect on the monsoon from a world away.

During an El Niño event, the Walker circulation weakens, causing a cascade of events. Local fishermen notice far fewer fish in their catches. As the transport of water across the equatorial Pacific Ocean weakens, leading to less upwelling in the eastern ocean, less cold water is added to the ocean off the coast of Peru. This creates a current of warmer water as the local sea surface temperature increases dramatically, causing fish to migrate south to colder waters instead.

This warm current is the El Niño current noticed by Spanish colonisers. The warmer waters heat the atmosphere over the eastern Pacific, leading to more atmospheric convection, lowering the air pressure and weakening the Walker circulation even further. The conveyor belt of heat and moisture slows to a crawl. Instead of making it to the other side of the ocean, rain now falls in the mid-Pacific, leading to droughts in Australia and Indonesia, dependent on rain brought by the Walker circulation.[19] Heat and moisture now don't flow as readily over into the Indian Ocean, reducing the temperature gradient between the land and the ocean, and thus weakening the monsoon winds. As a result of the air circulation weakening over the Pacific, the monsoon rains arrive late in India. In years with strong El Niño events, the monsoon can fail altogether, with devastating consequences, as we have seen. While El Niño doesn't always cause failures in the Indian monsoon, when the monsoon has failed, it has *always* been accompanied by an El Niño event.[20]

For completeness, there are events associated with the Walker circulation being anomalously strong, called La Niña conditions. A stronger than normal Walker circulation leads to more transport of water across the Pacific, colder water off the coast of South America, more precipitation in the western Pacific, and a stronger monsoon. These La Niña events tend to receive less attention, as they are not much distinguishable from normal weather, and have far fewer significant consequences.

I don't want to go into much more detail on oceans and their dynamics, partly because it is not my area of expertise, but more because if I kept filling in details on systems that influence the atmosphere, this book could go on forever! The atmosphere is a mediator of influences from a huge number of systems, and yet it is, as we have seen, a hugely complex system in its own right. ENSO is a spectacular duet between the Earth's oceans and its atmosphere, singing a melody that can devastate the local environment, with harmonics stretching over the entire planet.

Of relevance to our story here is the link between oceans and atmospheres, primarily through two mechanisms: moisture and temperature. As Brian Fagan details in *Floods, Famines, and Emperors*,[21] via these mechanisms El Niño and the Southern Oscillation have had enormous historical influence. This sloshing of warm water in the Pacific Ocean led to the fall of the Old Kingdom in Ancient Egypt, droughts that devastated the Mayan Empire, and shaped ancient civilisations from the Moche in Peru to the Anasazi in the American south-west, and, of course, civilisations throughout time in the Indian subcontinent.

These cultures could never have imagined that variations in an ocean current off South America could set off a cascade of global events, sending the atmospheric giant's footsteps crashing down from afar. With the benefit of modern analysis and modern data networks, scientists are now able to unpick signals from the interacting, interlacing fields of atmospheric and oceanic variability, spinning predictive gold from the straw of raw, global data. Atmospheric science allows us to predict the future using centuries of observations and theory, something that would have seemed like pure witchcraft to the people of the Mayan Empire or the Middle Kingdom of Egypt.

The state of Pacific water, being in El Niño or La Niña conditions, has consequences beyond the Indian monsoon, as far away as the mid-latitude jet stream. In Western Europe, local weather is manipulated by the jet stream, which is itself manipulated by global connections such as ENSO. But this is just one of a huge number of connections, with European weather ultimately forced by factors as diverse as Arctic ice cover, the wind in the tropical stratosphere, and rainfall patterns in the Indian Ocean. The atmosphere behaves on local, regional, and global scales, and these are all connected in sometimes surprising ways.

Meteorologists seem, then, to have the grand information pathways of the atmosphere mapped out, with one event linked to another, even a world away. The complexity is mind-boggling, but clearly understood.

So why is it that weather forecasts are still so unreliable?

CHAPTER 7

FORECAST

Tropical cyclones are some of the most well-known, devastating phenomena in the atmosphere. Depending on where you live, you may refer to them as hurricanes, cyclones, or typhoons, though all of these names refer to the same weather phenomenon. Air is heated by the tropical ocean underneath, causing extensive convection. As air rushes in to replace the partial vacuum left behind by the convection, it starts to rotate due to the Coriolis effect. The result is a warm, low-pressure system surrounded by rapidly rotating air. If this weather system forms over the tropical Atlantic Ocean, as around ten such systems do per year, and if the surrounding winds reach 120 kph (75 mph), as they do in around six per year, we call them hurricanes.[1]

In September 1985 the Gulf Coast of the United States was hit by one of these storms, named Hurricane Elena.* Despite the

---

* Hurricanes, and large Atlantic storms more generally, are named by the United States National Hurricane Center using one of six alphabetical lists of male and female names. Hurricane Elena was the fifth storm named in 1985, the fourth being Hurricane Danny and the sixth being Tropical Storm Fabian. Storms that are particularly devastating typically have their names retired; so, for example, there have been no Hurricane Elenas since 1985 and no Hurricane Katrinas since 2006. The fifth and eleventh names in the relevant lists now read Elsa and Katia respectively.

availability of vast quantities of data, supercomputers, and thousands of professional meteorologists, Hurricane Elena behaved in an entirely unexpected way. Rather than progress inland, as was forecast, the hurricane remained nearly stationary off the west coast of Florida for some forty-eight hours, devastating local communities. Beaches were severely eroded, fisheries were destroyed, and thousands were left homeless or unemployed. Meteorologists were confounded by the storm, advising the government to issue several evacuation orders over a wide area, trying to limit loss of life. Yet Hurricane Elena seemed impossible to forecast, eventually tracking inland and causing $1.3bn of damage, including damage to 13,000 homes, and the deaths of nine individuals.[2]

How was it that, despite centuries of scientific progress and some of the best meteorologists alive, a colossal hurricane could not be predicted? For that matter, why are weather forecasts both large and small so frequently incorrect?

To answer that question, we need to look at how we are able to forecast the weather at all.

William Ferrel's 1858 paper concludes with the line: 'We hope to be able to give a complete application of these principles to the theory of storms at some future time.' Presumably, what Ferrel meant by this was an application of physics equations to explain the behaviour of storms, and perhaps weather phenomena more broadly. Describing and explaining what is going on in the natural world, but not *predicting*. Meteorology as a science was at the time undergoing an explosion of interest and development, but the concept of prediction was just on the horizon.

The accelerating pace of meteorological observations in the eighteenth century continued into the nineteenth, critically aided by the invention – independently by several men in the

1830s and 1840s – of the telegraph. Suddenly, meteorological observations could be taken at huge distances apart and easily collated in a central dataset, nearly in real time. This potential was immediately recognised, and weather observation networks were established in Europe and North America. One such was created by none other than James Glaisher, our brave stratospheric aeronaut. He established a network of volunteers across the United Kingdom, who made weather observations every day at 9 a.m., telegraphing their findings to Glaisher at Greenwich Observatory in London.[3] In the United States, Joseph Henry (1797–1878), the Secretary of the Smithsonian Institution in Washington, collated weather information from across the country.[4] This marked a significant shift in data collection: not merely doing so incidentally, alongside some other activity such as running a colonial empire, but going out into the field and deliberately measuring characteristics of the physical environment. This formed part of what came to be known as Humboldtian science, named after German polymath Alexander von Humboldt (1769–1859), which came to define the Earth sciences in the nineteenth century.[5]

Originally somewhat haphazard, these meteorological observations would eventually be formalised into government offices, often as part of national governments. The first such government meteorological office was founded in Austria in 1851, swiftly followed by the UK Meteorological Office in 1854. This marked another significant shift in the history of meteorology. As with other fields of study, the nineteenth century saw the subject move away from folklore and individualistic approaches to quantitative data, with research conducted in large organisations. As we have already discussed, meteorology is perhaps unique in the nature of the data required to advance the field. An individual could never hope to collect global data sufficient to understand, say, the trade winds. A government office or university, by organising many individuals to work together, could answer questions about the atmosphere that an independent scholar simply could not. It should be noted, however, that, much as in other fields such as chemistry and geology, this institutionalisation of atmospheric science led to the exclusion of those who were simply not allowed to participate in large, scientific organisations. With this shift, for over a century women and people of colour all but disappear from the history of our atmosphere.

These governmental meteorological departments were intended to be data-gathering organisations, perhaps of some use to farmers in planning when to harvest. Some had more lofty ambitions, however. In a debate in the British parliament, one speaker suggested that by funding a national meteorological office, perhaps 'we might know in this metropolis the condition of the weather twenty-four hours beforehand'. Clearly, though, this was a minority view, as the official record shows the member of parliament ceased his speech due to the uproar of laughter his statement had caused.[6] Fortunately for the development of meteorology, there were at least a few who thought there just might be something in the idea of predicting, rather than merely describing, the weather. Perhaps the most important of these pioneers was the fascinating, tragic character of Robert FitzRoy (1805–65).

Best known as the captain of the expedition that carried Charles Darwin (1802–82) around the world, FitzRoy was also the founding figure of what we consider modern meteorology. Having excelled at the Royal Naval College at Portsmouth (now at Dartmouth), FitzRoy had a catastrophic first voyage in command. HMS *Beagle* hit a patch of rough weather and nearly capsized, losing two men in the process.[7] Barometers had been on board and indicated the falling pressure associated with a storm, but FitzRoy had not heeded their warning. He clearly learned from the experience. When he captained HMS *Beagle* again, this time on its five-year round-the-world adventure that led Darwin to formulate his theory of natural selection, the ship survived the most severe weather without the loss of anyone on board. Over five years, it never even once sustained any significant damage. This was a man who trusted his scientific instruments, and believed they could predict future conditions if read with enough skill.

After his most famous voyage, FitzRoy had a remarkable career. He served as a member of parliament, was the second governor of New Zealand, and in 1851 was elected to the Royal Society to head up a new department that collected weather data at sea. FitzRoy immediately saw the potential of the new data collection networks. If one instrument, read with skill, could predict future conditions, what could hundreds of observations accomplish? The quantity of data now at FitzRoy's fingertips – streaming in from ships, but also from coastal and inland weather stations – allowed him to construct what we now call *synoptic charts*.* These are maps of weather variables such as air pressure, temperature, and humidity plotted over a large geographic area, such as the British Isles, at a given time. Essentially, a snapshot of the atmosphere over a large area. FitzRoy studied the changes in weather across the British Isles using these charts and identified patterns that came up time and time again. He became increasingly convinced that the changes in weather at a given location could be predicted by interpreting what the atmosphere was doing at a large number of other, spaced-out observation stations. It was not long before FitzRoy

* Derived from being a *synopsis* of the state of the atmosphere, coming from the Ancient Greek words for 'together' and 'seeing'.

was drawing up synoptic charts of what he thought the weather would be doing in the near future. As early as 1842, he believed that if warnings of coming storms and other extreme weather could be issued to seafarers, many shipwrecks could be avoided and many more lives saved. His name for these predictions: *forecasts*.

Across Western Europe, other early meteorologists were thinking along similar lines. Since a disastrous storm that wrecked a dozen French warships in the Crimean War in 1854, the French government had established a network of meteorological stations so that seafarers could be forewarned of future storms. The Dutch meteorological institute was also exploring a similar storm-warning system. In 1859 a huge storm hit the British Isles, sinking a well-known steam clipper called the *Royal Charter*, with an estimated death toll of around 800 people.[8] This event catalysed FitzRoy into action, who was convinced that storm warnings could save lives. But he was not content with simply using the rapidly developing telegraph network to warn of storms already on the synoptic charts. He wanted to use his new forecasting method to predict how the weather would change. Writing in 1860, FitzRoy justified this strategy: 'It had been ascertained that atmospheric changes on an extensive scale were not sudden, and that premonitions were more than a day in advance, sometimes several days.'[9] A year later, in 1861, the first daily weather forecasts in the world, though vague and qualitative, were published in *The Times* of London by FitzRoy.

As a side note, the idea that this was even possible was so radical that it was, in the strictest interpretation of the law, illegal. The UK Witchcraft Act of 1735 defined claims to be able to predict the future as characteristic of witchcraft.[10] Until the repeal of the Act (in 1953!), all weather forecasters in the United Kingdom

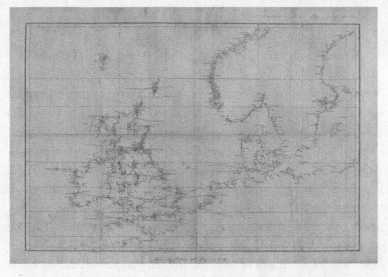

Figure 8: *An example synoptic chart constructed by Fitzroy, depicting the immense storm of 26 October 1859, in which the* Royal Charter *was sunk.*

were in the strictest interpretation of the law theoretically guilty of practising witchcraft. The underlying technology was so new and so unbelievable that its repercussions fell outside of the law, which took nearly a century to catch up.

Sadly, FitzRoy did not meet a happy end. The system of storm warnings he introduced was met with opposition by fishing fleet owners, and the very idea of weather forecasting was likewise consistently opposed. The latter, only enabled by the latest communication technology collating widespread data in real time, was so fantastical as to be unbelievable, and was seen, at the very least, as untrustworthy by the public. Because of these factors, financial support was not forthcoming for FitzRoy's forecasting work. He personally expended the equivalent of £400,000 in modern money on his weather-forecasting endeavours.

More troubling to him was an old ghost: his role in enabling Charles Darwin to construct his theory of evolution. Darwin's theory went directly against what FitzRoy – a devout Christian – saw as biblical truth, and his inadvertent support for Darwin's work on their shared voyage caused FitzRoy the 'acutest pain'. FitzRoy fell into a deep depression. His financial woes, combined with the moral weight of trying to save lives at sea in the face of great opposition, and the blame he laid on himself for eroding trust in biblical truth, became too much to bear. On 30 April 1865 he took his own life. FitzRoy was a transitional figure in the Victorian scientific scene, a man with one foot in the past and another in a future that few others could see. Today he is remembered as the man who founded weather forecasting, and who has helped to save tens of thousands of lives from storms and tempests, droughts and floods, and the extreme attentions of our atmosphere the world over.

FitzRoy very much drew on qualitative observations to make predictions about the future. As we have already seen, in tandem with this approach, the theoretical side of meteorology was also developing. William Ferrel, in particular, paved the way for the analysis of weather systems using physics and mathematics. At the turn of the twentieth century, these two branches combined. As early as 1890, American meteorologist and contemporary of Ferrel, Cleveland Abbe (1838–1916) proposed that meteorology was, in truth, simply the application of the laws of physics involving fluids and thermodynamics to the atmosphere.[11]

We can crudely approximate how we might do this. In previous chapters we have discussed how changes in pressure over large distances – horizontal pressure gradients – lead to wind. If we measured the air pressure at two locations, one, say, X miles east of the other, and called these measured air pressures

p_{west} and p_{east}, we could approximate that the east–west wind between the two, $wind_{EW}$, is something like

$$\frac{p_{east} - p_{west}}{X} \propto wind_{EW}$$

So, the greater the pressure difference between our two locations, the stronger the east–west wind; and the further apart the two locations are from one another, the weaker is the wind associated with a given pressure difference. The symbol in the middle of the equation simply means that one side of it is proportional to the other, rather than exactly equal.

What we have calculated is called the *pressure gradient force*: what fundamentally drives flow in the atmosphere. Except, of course, we're neglecting the effect of the Coriolis force. If we balance our pressure gradient force with the Coriolis force, saying that any pressure gradient is exactly matched by a Coriolis force from the resulting motion, we get a pretty good approximation of how the atmosphere behaves. This is technically called the geostrophic approximation,* with the wind calculated from making this approximation known as the *geostrophic wind*. We can crudely incorporate this into our approximated wind with an equation like

$$\frac{p_{east} - p_{west}}{X} \propto f \times wind_{NS}$$

where the wind is now the north–south wind, having been deflected, and we've also introduced a new term, f, to represent the Coriolis force at the location we're considering. This term is large over the poles and small near the equator, positive in one hemisphere and negative in the other, to represent the rotation of the Earth.

---

* From the Ancient Greek relating to the Earth – *geo* – and that relating to twisting/turning – *strophe* (as in the *troposphere*).

Recalling our discussion of fields in atmospheric science, and using information about how the pressure field varies horizontally, we can calculate an east–west velocity field, and a north–south velocity field, over a large area. Together, these two components tell us the wind speed and direction at any location. When viewed on large scales, such as the circulation around an entire hemisphere, or around a large hurricane, the calculated geostrophic wind matches observed wind reasonably closely. One way to think about geostrophic wind is that it flows along the *isobars* – lines indicating constant pressure – on synoptic charts. Around significant, large pressure gradients, such as the depressions found at the hearts of extra-tropical storms, the air does indeed follow the pressure contours.

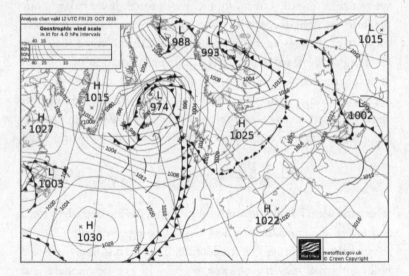

Figure 9: *A modern synoptic chart. The thin lines are isobars – lines of constant air pressure – while the thick lines are weather fronts, the boundaries between different air masses. The shapes on the weather fronts indicate their type, e.g. a cold front is denoted by triangles while a warm front is denoted by semicircles.*

So, from widespread measurements of the air pressure, an east–west velocity field and a north–south velocity field can be calculated. Using these velocities, future changes in the pressure field can be predicted via a process call *advection*. Advection is, to put it crudely, the process of a fluid carrying stuff to places. Wind, the flow of atmospheric fluid, carries moisture from oceans over the land. It carries heat from the tropics to the polar regions. It carries clouds in swirling patterns around storms. Advection is the fundamental process of how the atmosphere – and indeed all fluids – transports moisture and temperature and other quantities from place to place. Even linear and angular momentum can be transported in this way, influencing the velocity field that itself gave rise to the advection. Having calculated a velocity field, which could be geostrophic or more complex, calculating the advection field allows scientists to 'turn the crank' into the future, predicting how atmospheric fields will change in the future as a result of the calculated velocity field.

In 1904, the Norwegian meteorologist Vilhelm Bjerknes (1862–1951) took this rough argument and made it much more concrete, arguing that weather forecasting should be possible using mathematical methods. If that name sounds familiar, Vilhelm was the father of Jacob Bjerknes, the man who unpicked the duet between El Niño and the Southern Oscillation in the previous chapter – they formed quite the academic dynasty!*

---

* It wasn't just the father and son team either! Vilhelm's father Carl Anton did extensive theoretical work on fluids and the similarities between their mathematics and those of electric and magnetic fields. He never experimentally tested his theories, however, and so at the age of just seventeen Vilhelm took it upon himself to devise a series of experiments by which his father's theories could be verified. He was successful, and his demonstrations using a bath full of syrup were exhibited at the International Exhibition in France in 1881 and were of huge interest to scientists.

Combining the work of pioneers like Ferrel with the more theoretical work of Newton, Coriolis, Euler, and many others, Bjerknes proposed a simple system of equations to describe the flow of the atmosphere. These would later – cruelly, according to undergraduates – be called the *primitive equations*. In truth, while these equations look incomprehensible to someone without mathematical training, they are actually very simple at heart.[12] Bjerknes basically argued that you needed to follow four steps to predict what the atmosphere will do next:

1 Take measurements of atmospheric fields such as pressure and temperature, providing you with the initial state of the atmosphere.
2 Calculate expected changes in the velocity fields using your observed initial state and the primitive equations.
3 Using the updated velocity fields, calculated in the previous step, work out the expected changes of fields such as temperature and humidity due to advection.
4 Repeat steps 2–4.

$$\frac{\partial u}{\partial t} + u\frac{\partial u}{\partial x} + v\frac{\partial u}{\partial y} - fv = -\frac{1}{\rho}\frac{\partial p}{\partial x} + F_x$$

$$\frac{\partial v}{\partial t} + u\frac{\partial v}{\partial x} + v\frac{\partial v}{\partial y} + fu = -\frac{1}{\rho}\frac{\partial p}{\partial y} + F_y$$

$$\frac{\partial p}{\partial z} = -\rho g$$

$$\frac{\partial \theta}{\partial t} + u\frac{\partial \theta}{\partial x} + v\frac{\partial \theta}{\partial y} = 0$$

$$\frac{\partial \rho}{\partial t} + u\frac{\partial \rho}{\partial x} + v\frac{\partial \rho}{\partial y} = -\rho\left(\frac{\partial u}{\partial x} + \frac{\partial v}{\partial y}\right)$$

Figure 10: *One formulation of the primitive equations (see page 130 for further description).*

At its core, that's it! It simply took three centuries of development of physics and mathematics to get to that point, and roughly the same amount of time to develop the tools and observation networks necessary to work out that the atmospheric flow was indeed that simple. Of course, I'm playing down the complexity of the real science, and there are many, many, many additional factors that go into modern equations describing how the atmosphere behaves. But at their heart, those modern equations are still the same as those developed by Bjerknes. His ideas were so influential that they evolved into the *Bergen School of Meteorology*, a school of thought used in basically all of modern atmospheric physics, and particularly adopted by influential early twentieth-century Scandinavian scientists such as Carl-Gustaf Rossby.[13] These scientists used the rapidly developing observation networks, considerably expanded during the two World Wars, to construct what we would now recognise as modern meteorology.[14]

On large scales, meteorology is principally concerned with calculating the evolution of different *air masses* and the boundaries between them, known as *weather fronts*.[15] Air masses are essentially large blobs of air, typically on the scale of an entire country, defined by their temperature and their humidity. For example, continental polar air masses are cold and dry, while maritime tropical air masses are warm and moist. Underneath a given air mass, weather will be relatively consistent, with that weather dependent on the classification of the air mass. When one air mass meets another, however, significant changes in weather can occur at the boundary. These weather fronts, so named due to their resemblance to battle lines on the western front of the First World War, come in a few flavours, such as cold fronts, warm fronts, and occluded fronts. Each category of front represents a particular combination of atmospheric conditions. A cold front, for example, occurs when a cold air mass rolls over the landscape, displacing and uplifting warmer air. The uplift that occurs can produce sharp changes in weather, such as rapid cloud formation and heavy precipitation.

One of the beautiful things about atmospheric physics is that these small-scale changes along the weather front can also be described, and predicted, using the same equations as those describing continent-sized air masses. Interacting air masses behave as fluids, as do small-scale air parcels rising and falling at weather fronts. The difference between these two scales is what physical processes can be ignored in the equations describing their motion. At large scales, for example, friction between air masses and the ground can be safely ignored, while at small scales friction with the ground can be of absolute importance. Air flowing over terrain such as mountains or valleys produces totally different effects to air flowing over flat ground. The huge variety of cloud types that we observe, from cumulus to cirrus to nimbostratus, are made possible by specific small-scale processes, including uplift by terrain, but also by other factors such as local wind shear, temperature gradient, and humidity.

The physics of the atmosphere may be universal, but that universality is overwhelming. In order to make sense of it, scientists must pick their battles, ignoring complexity that can be safely ignored, and focusing on the physical processes that control the behaviour we observe. Over the past two centuries, our understanding has evolved to the point that the future has opened up to us.

Meteorologists use the concept of air masses and fronts to explain and predict how weather changes over time both qualitatively and quantitatively. *Numerical weather prediction* – meaning using the equations of the Bergen School to predict the future state of the atmosphere given current measurements – by computers allows for quantitative prediction, while a trained meteorologist can view a synoptic chart and, seeing the configuration of fronts, make a qualitative prediction of the next few hours. Fronts break down the continuum of interconnected atmospheric fields to a manageable set of interactions between discrete entities, allowing our human brains to grasp, and predict, the complexities of the atmospheric giant.

A computer, of course, does not need this conceptual shorthand. Instead, a computer model used for quantitative weather prediction stores the information describing an atmospheric field, such as temperature or humidity, in a grid. The *grid size* or *resolution* determines how many data points are used to describe the field – if we are considering a field on a global scale, a resolution of 2.5° might be used, indicating that a value is stored at every 2.5° in longitude and latitude. The higher the resolution, the more accurately the continuous field is described, though this makes computation slower. Early computer models had very low resolution, limited by the size of their memory, while modern computer models can have extremely high resolutions. The UK Met Office uses a resolution of just 1500 metres when running forecasts over the British Isles.[16]

It should be noted, however, that numerical weather prediction predates the computer. Or, at least, the modern idea of the computer. A *computer* was, originally, someone who computed, who performed mathematical calculations over and over again, this tedious work often being undertaken by women, or teams of women. One male computer was Lewis Fry Richardson (1881–1953) – a man of wide-ranging interests, but whose Quaker beliefs led him to be a conscientious objector in the First World War and so disqualified him from holding any academic post. In 1922 he published *Weather Prediction by*

Numerical Process,[17] a strikingly original book sadly hamstrung by the limitations of his time. In it, Richardson presented a systematic method of forecasting the weather using dynamical equations, almost exactly as modern computer models do. He also demonstrated his method by calculating such a forecast *by hand*. That is, instead of getting an electronic computer to perform the calculations of changes in air pressure, temperature, and the like for every grid point in his forecast area, he worked out the predicted changes over eight hours using pencil and paper. The monumental effort took him six weeks. Sadly, though perhaps not surprisingly, his forecast failed dramatically, predicting a huge change of 145hPa in pressure at the surface, a number around one hundred times too large! According to later analysis by meteorologist Peter Lynch, however, this was probably more to do with how Richardson prepared his data than how he calculated the changes. Had this been accounted for, Richardson would likely have worked out a reasonably accurate weather forecast for a day by hand – a truly remarkable achievement!

As it was, the idea of numerical weather prediction seemed fanciful, doomed to fail by the sheer number of calculations required. It took three decades for technology to catch up to the dreams of the Bergen School and Richardson. In the 1950s, the first general-purpose electronic computers such as the Ferranti Mk I and the UNIVAC I were released, building on rapid advances in digital computing during the Second World War. These early computers were enthusiastically taken up by scientists of all stripes, and in 1950 the first digital meteorological forecast was made by a superstar team including meteorologist Jule Charney (1917–81) and mathematician John von Neumann (1903–57). The feeling at the time was that science and technology could accomplish anything, and with the hitherto unimaginable number-crunching power of digital computers, scientists tried their hands at problems previously considered unsolvable. Granted, the superstar group's prediction for the following twenty-four hours of weather took nearly twenty-four hours to produce, and was extremely limited in detail, but the potential was clear.[18] Perhaps science could indeed predict what the weather was going to do next?

During the 1950s, the atmosphere was simulated with increasing complexity on computers, and various questions about the atmosphere investigated besides predicting the weather. Many of these were problems that had previously confounded researchers by being just too complex to solve using pencil-and-paper calculations. One such problem was how overturning circulations in the atmosphere such as the Hadley cell organise themselves – how big the temperature difference north and south of circulations becomes, how strong convection is in the atmosphere, and how high the circulations reach. In 1961, a quiet, unassuming meteorologist named Edward Norton Lorenz (1917–2008) was investigating this problem using a highly simplified set of equations.[19] Perhaps I have been indoctrinated by my time in academia, but I find the equations he derived really rather beautiful. Using just seven variables and in three elegant lines, the behaviour of a hugely complicated system the size of a planet is captured.

Assisted by legendary computer scientist Margaret Hamilton (1936–), then just twenty-five,* Lorenz was examining this problem when he decided to repeat some calculations in order to

---

* Hamilton went on to have a trailblazing career, so fundamental to software engineering that she even gave the field its name. Among other accomplishments, she was director of the Software Engineering Division of the Instrumentation Laboratory at MIT, developed software for the Apollo missions to the Moon, and has her own LEGO minifigure.

more closely examine what was going on. Writing much later in *The Essence of Chaos*,[20] he describes what happened next:

> I stopped the computer, typed in a line of numbers that had come out of the printer a little earlier, and started it back up. I went to the lobby to have a cup of coffee and came back an hour later, during which the computer had simulated about two months of weather. The numbers coming out of the printer had nothing to do with the previous ones.

This was surprising! Lorenz had taken the output from an earlier simulation of the atmosphere to use as a starting point – what in science we call an *initial condition* – for a new, identical simulation to repeat a section of interest. Yet the results he'd got from the computer from this second simulation were totally different to his previous results. This shouldn't have happened. These equations have no randomness in them: they are what we refer to as *deterministic*. Much like Newton's laws of mechanics, if you know the initial condition – the lie of the land when you start out – then by applying the same equations you should get the same result every single time.

Digital computers were still new and subject to mechanical error, however, and Lorenz thought this must have been the cause. He wrote:

> I immediately thought one of the [vacuum] tubes had deteriorated, or that the computer had had some other sort of breakdown, which was not infrequent, but before I called the technicians, I decided to find out where the problem was, knowing that that would speed up the repairs. Instead of a sudden interruption, I found that the new values repeated the previous ones at first, but soon began to differ by one or more units in the final decimal, then in the previous one, and then

the one before that. In fact, the differences doubled in size more-or-less constantly every four days until any resemblance to the original figures disappeared at some point during the second month.

Had Lorenz discovered a set of deterministic equations that were *not* actually deterministic? Not quite. The equations were still deterministic: if you put the same numbers into them time after time, you would get the same results time after time. The problem was that Lorenz *wasn't using the same numbers*. He had taken his initial condition from the printout of the model's variables, taking numbers off the page such as $x = 0.506$ and $y = 0.193$. The printer would print numbers to three decimal places. The model, however, didn't work to three decimal places – it worked to six. While the printer might say the model calculated y to be 0.193, the model *actually* calculated it to be 0.193208. Therefore, Lorenz wasn't restarting the model with *exactly* the same initial conditions: he was restarting it with *almost exactly* the same initial conditions. At first this difference proved negligible, as you might expect. But over time, the two models with imperceptibly different initial conditions began to diverge from each other more and more, until eventually they bore no resemblance whatsoever.

Lorenz went on to detail this phenomenon (in surprisingly readable prose; it's worth giving it a read!) in one of the great physics papers of the twentieth century, 'Deterministic Nonperiodic Flow'.[21] Due to its wide-ranging implications, Lorenz's discovery birthed an entirely new academic field. For over a decade, scientists in the United States and Soviet Union investigated similar phenomena in pure mathematics, economics, biology, physics and other fields, but did so without a name for what they studied. It was only in 1975 that mathematician James

Yorke (1941–) coined the phrase that launched a thousand popular science books: *chaos theory*.[*]

Chaos theory, at its heart, examines dynamical systems that may appear to behave randomly but are in fact governed by deterministic equations, like Newton's laws of motion. These equations are combined in such a way, however, that outcomes are extremely sensitive to initial conditions. Lorenz neatly summarised it as: 'Chaos: when the present determines the future, but the approximate present does not approximately determine the future.'[22] Weather is still perhaps the most famous example of a chaotic system. The weather we experience on a given day may seem random, and for the longest time was believed to be all but so. Yet the weather is in fact governed by a small set of equations with definite outcomes, just ones that are very sensitive to the initial conditions put into them.

The chaotic nature of the atmosphere is why weather forecasts are often inaccurate, as was the case with Hurricane Elena in 1985. If information about the initial state of the atmosphere is incomplete, accurately predicting even a short distance into the future can be next to impossible. While scientists have access to data from a huge number of locations, indeed, globally, thanks to satellites, the fact remains that computer models only operate at a certain resolution, meaning that only so much data can be incorporated into predictions. Even if data was available at one-centimetre increments across the face of the Earth, only so many of these data points could be used to make a prediction.

---

[*] This was not the first time that chaos theory had been discovered! That was done by polymath Henri Poincaré (1854–1912) in the 1880s and mathematically described by Andrey Nikolaevich Kolmogorov (1903–87) (see C. Oestreicher, 'A History of Chaos Theory', *Dialogues in Clinical Neuroscience*, vol. 9, no. 3 (2007), pp. 279–89), but their theories never caught on.

As if that wasn't bad enough, each data point also has a (small) uncertainty with it. Consider a thermometer – it may register a temperature as being 20.1 °C, but because of rounding that could be any value between 20.05 °C and 20.15 °C. That may sound like a small error, but when multiplied across thousands or tens of thousands of measurements, it adds up fast. In 1985, forecasters had a large amount of data at their disposal, from ground-based sources and satellites, but were limited by the (still high) accuracy of that data and the (relatively low) resolution of their predictive models.

It's possible for scientists to get around these difficulties by running *ensemble forecasts*. This means predicting the weather several times, but initialising each set of calculations with a slightly different set of initial conditions. One prediction may take the upper end of the uncertainty in the temperature field as its initial condition, for example, while another may take the lower end of the uncertainty in the temperature field as its initial condition. Each of these forecasts will produce different end results, which may be very similar, or wildly different. Taking the average of all these forecasts produces one, more statistically robust, forecast.

Most of the time this is sufficient to circumvent the chaotic difficulties of the atmosphere, and predict the weather up to a week ahead of it happening. The UK Met Office predicts the following day's temperature to within 2 °C approximately 92 per cent of the time. Additionally, thanks to improvements in methodology, computing hardware, and satellite technology, this accuracy has increased over the past century. The Met Office can predict weather four days ahead to the same accuracy now that it was able to predict one day ahead in 1980.[23] Sometimes, however, particular conditions lead the atmosphere to be particularly unpredictable, as happened with Hurricane

Elena and as will certainly happen in the future. The inherently chaotic nature of the atmosphere means that even with the most widespread measurements possible, the highest-resolution computer model, and the most accurate instruments, the small error bars present in the initial conditions of the atmosphere eventually propagate to huge errors in forecasts far into the future. Practically, this means that science can predict the weather really quite accurately for about ten days into the future before the errors become unacceptably large. This isn't a flaw in how we represent the atmosphere, in our equations or our supercomputers. It's a fundamental limitation on predicting the future of a chaotic system.

However, for up to two weeks from the present, three centuries of scientific progress allow us to perform a miracle and gaze into the weather yet to come.

Figure 10: *Here u and v are, respectively, the east-west and north-south components of wind, while x, y, and z are, respectively, the distance travelled east, north, and up from some location. Note that in this formulation, the vertical velocity is assumed to be negligible. f is the Coriolis parameter, p is the air density, p is the air pressure, Fx and Fy are frictional forces in the x and y directions, and θ is the potential temperature (this is the temperature an air parcel would have if you dragged it down to the surface). The first two equations describe how an air parcel accelerates in the east-west and north-south directions, while the third equation describes how pressure varies as you get higher in the atmosphere. The last two equations are book-keeping, making sure that energy and mass are conserved, respectively.*

CHAPTER 8

VORTEX

From 22 February to 5 March 2018, the British Isles shivered under Anticyclone Hartmut, better known as the 'Beast from the East'. For two weeks, frigid air transported all the way from Siberia sat over Europe, causing temperatures to fall to -14 °C (7 °F) in Scotland. This cold air mass was then struck in the west flank by a moisture-laden storm from the Atlantic, producing huge amounts of snow and ice. Ninety-five people lost their lives to the cold across Europe, due to road traffic collisions, falls on ice, and inadequate shelter. Altogether, the cold outbreak caused some £1.2bn of damage.[1]

Less than a year later, on 30 January 2019, it was colder in the city of Chicago, at latitude 42 °N, than at the south pole. In fact, it was colder than Mars.[2] The temperature in Chicago plunged to -30 °C (-22 °F), causing chaos, closing the local government, freezing water mains, and leaving people advised not to talk or to breathe too deeply outside, to protect their lungs from the cold.

Both of these events were caused by the same spectacular weather phenomenon, created by a true titan of the atmosphere. So far in this book we have stayed relatively close to the surface, looking at weather and dynamics in the bottom layer of the atmosphere, the troposphere. For one chapter, however, allow me to introduce you to the monsters of our upper atmosphere. This will include a bit of my own research, explaining how these titans reach down with icy fingers to shape our weather on the surface.

If you had lived in the east of England in 1915, you would have occasionally heard the big guns firing on the western front. The use of artillery in the First World War was of such an unprecedented scale that locals reported hearing summer battles as a constant, low rumble. Eventually, the guns would fall silent, and their sound faded to memory. However, in 1935, mathematician Francis John Welsh Whipple (1876–1943) delivered a lecture to the Royal Meteorological Society in London (of which he would later be president) titled: 'The Propagation of Sound to Great Distances'.[3] In the lecture he credits a Mr Miller Christy, who lived near Chelmsford, Essex, with keeping a diary in which he noted day by day if the sounds of the battlefields could be heard. Whipple analysed these diary entries and found something surprising. There was a striking seasonality to them – in summer, artillery could be clearly heard, while in winter it could not. Fighting died down in the winter months, but artillery was still used, and yet could not be heard at all.

As the fighting was taking place far to the east of Chelmsford, Whipple concluded that the sound was being carried westwards in summer. As sound waves inherit the velocity of any fluid layer they travel through, even if only briefly, he hypothesised that this carrying was done by some strong wind. The scientific consensus at the time was that a general east-to west circulation prevailed in the newly discovered stratosphere, moving opposite to the Earth's rotation. Such a circulation would cause sounds to be heard far further west if they were loud enough to reach, and reflect off, the tropopause, the boundary between the troposphere and the stratosphere. While this theory certainly explained the sound of artillery in France carrying to England in the summer months, it did not explain the quiet winters described in Mr Christy's diary. Something was awry. However, frustrated by a lack of data above 20 km in altitude – as the cutting-edge method of data collection at this time was the weather balloon – Whipple could only conjecture about what was going on.

He hypothesised that the winds in the stratosphere must reverse as summer changes to winter. With the wind blowing from west to east, the sound of gunfire would be carried further east from France, leaving England quiet. But what could be the cause of such a reversal? As we learned previously, the defining characteristic of the stratosphere is its air temperature increasing with altitude, leading to very different dynamics to those seen in the troposphere. In particular, the static stability of the stratosphere inhibits almost all vertical motion, causing air to flow horizontally instead, as if on a flat surface, in far larger patterns. With no vertical motion to disrupt them, immense atmospheric structures the size of continents can form. Instead of splintering into small storms, these titanic circulations grow larger and stronger over the course of the year. The only thing that checks their growth is the changing of the seasons.

At the equator there is precious little difference between the seasons – in fact, the very concept of summer and winter seasons only really make sense from the mid-latitudes upwards. Here, there is a marked difference between the warm, sunny summer and the cold, dark winter. But further than approximately 66.5° from the equator, in the Arctic and Antarctic circles, winter is a very different beast. Due to the Earth's tilt of approximately 23.5° between its axis of rotation and the orbital plane, these areas experience what's known as the *polar night*: periods when the Sun never rises. They also experience *midnight sun* in summer, a period when the Sun never sets. The difference, then, between summer and winter in the polar circles is vast, with all-day heating in the summer and all-day darkness in winter.

The polar night is bitterly cold: temperatures are often recorded as lower than -50 °C on the surface, and less than -80 °C in the middle atmosphere. From our discussion of the ideal gas law, we know that a decrease in temperature with no change in air density will result in a decrease in atmospheric pressure. And, as we also previously learned, when areas of low atmospheric pressure form on the spinning Earth, air will rotate around them in a cyclonic circulation. A vast, hemisphere-scale circulation therefore develops around the polar region in winter, covering an area the size of Asia. When summer arrives, the Sun rises on the polar region once more, and a huge amount of energy in the form of sunlight is added. Thanks to the absorption of UV radiation by ozone, this causes a spectacular rise in temperature, and the summertime polar stratosphere is warmer than the equatorial stratosphere. This destroys the wintertime circulation, and the polar regions are now home to higher air pressure than the low latitudes. As such, an anticyclonic system, spinning in the opposite direction, forms.

This explains the reports of artillery heard in England, and Whipple's conjectured reversal of stratospheric winds. When the seasons change and summer comes to the Arctic circle, the massive stratospheric circulation is disrupted and reversed. But we could ask a very reasonable question: why doesn't this happen closer to the ground too, in the troposphere? After all, the polar night is felt here too. Why don't we see this reversal of winds with the seasons as well?

We don't see such a dramatic shift in the tropospheric circulation for a couple of reasons. The main one – and the author feels a certain kinship here – is to do with the Earth's bulging midsection.

This is something that I previously mentioned only in passing: that the Earth is not in fact a sphere. The planet's rotation causes its rock and mantle to bulge slightly at the equator, making it technically an oblate spheroid.[4] If you were to place a tape measure around the Earth's equator and cut through all terrain to sea level it would measure the Earth's circumference to be about 40,075 km. Place the same tape measure with one end at the north pole, run it due south to the south pole, and then continue on the same path back to the north pole and you will find you have tape to spare. The *polar circumference* of Earth is only 40,008 km. That means there is a greater distance between you and the centre of the Earth when standing at the equator than when standing at the poles.

As gravitational acceleration is inversely proportional to the square of distance, this difference results in surface gravity being lower at lower latitudes. Combined with the centrifugal acceleration associated with the Earth's rotation, you quite literally weigh less when standing closer to the equator! In fact, an argument could be made for some Olympic records such as high jump to be made latitude specific – after all, an athlete jumping in London is straining against stronger gravity than an athlete in, say, Nairobi. The difference in surface gravity is at most half a per cent, but this can still be significant in competitive sports.[5]

The slightly reduced gravity at the equator has a much more noticeable effect on the atmosphere. Much like the planet underneath, it bulges outwards at lower latitudes. The troposphere extends much higher at the equator than it does at the poles. In low-latitude regions the troposphere ends at around 17 km, while in high-latitude regions it tops out at as little as 8 km. Between these two extremes the atmosphere gradually slopes. This is partly due to the reduced gravity at the equator, and partly down to the more extensive heating that the atmosphere over the equator receives. These two effects together cause a lopsided distribution of pressure in the atmosphere. Much as the Earth's surface bulges at the equator, so too do the pressure surfaces in the atmosphere.

Or, to view it another way, one kilometre above the north pole the air pressure is significantly lower than the air pressure one kilometre above the equator. Recall also that, due to their lower temperature, lower pressure already exists in polar regions. This means that as you get higher and higher above the ground at high latitudes, the pressure difference between the high- and low-latitude regions increases and increases. This increasing pressure gradient with altitude produces a *wind shear*, as the strength of the wind is proportional to the strength of the pressure gradient, and as the gradient gets steeper with altitude, the strength of the wind also increases. This is what is known as the *thermal wind*.[6]

As was originally conjectured, the normal state in the summer-time stratosphere is indeed for air to gently flow from east to west, the constant insolation (the technical name for energy coming in from the Sun) of the high latitudes causing air there to be slightly warmer than air at the same vertical level in the tropics. This temperature gradient causes air to drift equatorward and be deflected by the Coriolis effect into an easterly wind. As winter sets in across a hemisphere, however, causing the Sun to set for months at a time, temperatures plummet, and a temperature gradient is established in the opposite direction: now the polar air is significantly colder than air at the equator. This causes air in the stratosphere to rotate around the pole in the opposite direction. With the winter night bringing bitterly cold temperatures and low air pressure, the rotating air picks up speed and starts to dominate the circulation in the middle atmosphere as a strong westerly wind. The thermal wind effect causes this circulation to be much, much stronger in the stratosphere than in the troposphere, accelerating the wind to breakneck speeds.

The huge rotating circulation that results is the *stratospheric polar vortex*, the ghostly titan of the atmosphere. This is the culprit ultimately behind the frozen outbreaks of air in London and

Chicago. A polar vortex forms every year in whichever hemisphere is currently experiencing winter, stretching from approximately 15 km to more than 50 km above the surface. The stratospheric polar vortex is vast. The largest tropical storm ever observed, Typhoon Tip, was over 2,000 km across and had wind speeds peaking at 300 kph (190 mph).[7] By contrast, the polar vortex is around *6,000 km* from edge to edge – quite literally the size of a continent. The southern vortex is significantly stronger and more uniform than the vortex that forms in northern hemisphere winter, for reasons that we'll shortly come back to, with wind speeds around 300 kph. A typical day in the southern stratospheric polar vortex features wind speeds similar to the strongest winds ever recorded in a typhoon or hurricane! The northern vortex averages wind speeds around 200 kph (125 mph), but this varies significantly.

It should be pointed out that often scientists abbreviate the term 'stratospheric polar vortex' to just 'polar vortex', and this has led to some confusion in the public, who often mis-attribute the term to the jet stream. While some literature does discuss the 'tropospheric polar vortex', and the jet stream and stratospheric polar vortex are closely linked, they are still distinct phenomena. When I refer to 'the vortex' in this chapter I am explicitly referring to the stratospheric polar vortex.

While the vortex is truly immense in both size and velocity, remember that it is also ghostly – the air density in the stratosphere is between one-hundredth and one-thousandth of what it is at the surface. The air here moves quickly, but it does not carry with it much weight. So how could such a thing influence the weather at the surface, enough to bring polar temperatures to Chicago? For several decades after the discovery of the stratosphere, the prevailing wisdom was indeed that the air at such altitude was too thin to have significant impacts on the surface.

Compared to the dense troposphere, the stratosphere was the tall weakling on the atmospheric playground, pushed around and never in control. Other than the seasonal reversal of winds, not a huge amount was believed to happen in this region. Over time this view came to be challenged, however, and would eventually be shattered by an atmospheric explosion.

In 1952, Richard Scherhag (1907–70), professor at the Freie Universität Berlin, published a paper with the devastatingly wonderful title of 'Die explosionsartige Stratosphärenerwärmung des Spät-winters 1951/52', or, in English, 'The Explosive Stratospheric Warming of Late Winter 1951/52'.[8] Using data from weather balloons, Scherhag observed the air temperature around 30 km above Berlin skyrocket by some 40 °C in just a few days. It must have been an amazingly exciting discovery – the speed of the observed warming in the atmosphere was so unprecedented that the label 'explosive' was absolutely justified, though sadly it never caught on in the scientific literature. Originally referred to as the 'Berlin warming', the phenomenon came to be known as an example of a 'sudden stratospheric warming', or SSW for short. As would be discovered with subsequent observations, during these SSW events the air temperature in the middle atmosphere would increase by as much as 60 °C in just a few days, and they were not restricted to central Europe – the entire region of the stratospheric polar vortex was affected. Sudden warming events were observed all over the Arctic sky, occurring on average six times every decade.

The explosive warming was the standout feature, the poster boy of SSWs, and as such got top billing in the name. But, to me, the far more shocking thing about these events is what happens to the vortex itself. When an SSW occurs, the polar vortex – the continent-sized doughnut of air spinning at 200 kph – tears itself apart. The immense warming fundamentally destabilises the hemispheric temperature gradient that drove the vortex, resulting in its destruction. Sometimes this involves the circulation making a sudden journey equatorward, eventually hitting such warm air that it cannot sustain itself, while sometimes the circulation stays in place but tears itself into two smaller vortices. These two circulations then fight it out, always resulting in the total destruction of both. Either way, a rapidly spinning mass of air the size of Asia, over the course of perhaps a week, utterly destroys itself. It really cannot be overstated how dramatic these events are.

With the advent of satellite observations, we now have a much clearer picture of the lifecycle of the polar vortex and so can put Scherhag's discovery in more context. As we've already discussed, the vortex forms as winter sets in across a hemisphere. In the northern hemisphere this is typically around November. Through the winter the vortex maintains a strong westerly circulation, but as spring arrives sunlight creeps into the polar regions and causes them to warm, and the vortex gradually gets weaker and weaker.

The stratosphere's transition from wintertime regime to summertime regime is not as smooth, however. While in late autumn the vortex enters the scene slowly without much of a fuss, in spring it exits stage with a bang. Or, more precisely, an explosive warming. The vortex shreds itself in what is known as a *final warming*, to distinguish it from the sudden stratospheric warming that Scherhag discovered. This final warming occurs every year, and marks the end of the winter season. SSWs on the other hand can occur at any time in the winter, and are followed by the vortex pulling itself back together and resuming a westerly circulation over the course of a few weeks, before eventually succumbing to a final warming in spring.

I previously mentioned that the southern polar vortex is much stronger and more uniform than the northern polar vortex. In all our years of observations, the southern vortex has only experienced three SSWs (and one of those was very minor). There is a simple reason for this, and it involves the same factor behind the violent disruptions to the northern vortex: the fact that the Earth's southern hemisphere is mostly ocean.

This may seem like a very odd statement, but go and find a globe and you'll see what I mean. The northern hemisphere is jam-packed with both continents and oceans, but flip the world upside down and you'll see mostly ocean in the southern hemisphere. This lack of land/sea contrast means that the hemisphere has a comparatively *zonally symmetric* atmosphere. Along a given latitude line, you will find the atmosphere doing much the same thing.

Why is this important? Different materials possess different heat capacities – the exchange rate between energy and temperature change – and will therefore change temperature by different amounts even when heated by the same amount of sunlight. The ocean, for example, responds far more sluggishly to changes in solar energy, warming more slowly than the land in spring, and cooling more slowly than the land in autumn.

Different temperatures – such as over land and over oceans – along lines of constant latitude produce areas of high and low pressure, which we call *atmospheric waves*. These waves can move, or, more properly, propagate, both over the surface of the Earth and also vertically up or down through the atmosphere. We see this as areas of high and low pressure moving in tandem around a line of constant latitude, much as we see ocean waves moving together as peaks and troughs in the water. Ocean waves propagate, forced from above by the wind, until they reach a shore. When such a wave reaches land, it breaks. Doing so it transfers its momentum – the velocity and mass of the water that has had its movement interrupted by the land – into the shore. It does this because the wave has encountered a medium into which it is unable to propagate – solid land – yet its momentum must be conserved. As a result, crashing waves impart great forces onto the land, which cause the gradual erosion of solid rock into spectacular coastal landforms.

While the continental variation of atmospheric pressure across an entire planet may seem completely unlike these ocean waves, atmospheric waves are described by the same fluid dynamics and can therefore also 'break' in the same way when they encounter a region through which they can no longer propagate. However, for atmospheric waves, this doesn't take place where the ocean meets the land but instead where certain atmospheric conditions are no longer met. Specifically, this occurs at the point where they encounter the stratosphere and meet the polar vortex, in a region just outside the edge of the vortex called the *stratospheric surf zone.*[9] Here, atmospheric waves 'break' – or deposit their momentum, to describe it a little more formally – but instead of eroding the landscape, this deposition of momentum acts to slow the stratospheric polar vortex down. Due to a quirk of geometry, this deposition only ever acts to slow the vortex – it never accelerates the stratospheric flow. If the polar vortex is like a free-spinning bike wheel on its side, constantly accelerated by the equator-to-pole temperature gradient, the breaking of atmospheric waves acts as a brake on the wheel, preventing it from accelerating to ever greater speeds. The stronger the braking force, the slower the wheel ends up spinning.

As the northern hemisphere has a large amount of both land and sea – it is highly zonally asymmetric – it produces strong atmospheric waves, which act as a significant brake on the northern polar vortex when they encounter the stratosphere and deposit their momentum. By contrast, the southern hemisphere is largely zonally symmetric and so produces weaker atmospheric waves that act as a weaker brake on the southern vortex when they break in the southern hemisphere stratosphere. As a result, the southern vortex is much stronger than the northern vortex, even though they are both accelerated by the same temperature gradient.

But what does this have to do with sudden stratospheric warmings? I promised that the lack of zonal symmetry in the northern hemisphere also explained why these hugely violent events were more common here. Well, I somewhat over-simplified our discussion of atmospheric waves. Yes, they are driven by land–sea contrast and zonal asymmetry, but they are also driven by local conditions such as precipitation and heat transported from nearer the equator. This means the production of atmospheric waves in a given hemisphere is variable – depending on these extra factors, at a given time there can be lots of wave activity, or not very much. Just as the degree of zonal symmetry controls the braking of the polar vortex, this extra dash of variability acts to further control the strength of the polar vortex.

When the wave activity is particularly large, the brake on the vortex can become so powerful that the wheel stops spinning entirely, and the polar vortex careens to a halt.* As the northern hemisphere has a higher-than-average wave activity, when the additional factors like local precipitation are stronger than normal, this results in enough wave forcing to cause a sudden stratospheric warming. Compare this to the southern vortex, where the average wave forcing is lower, and the additional factors would have

---

* The mathematics of this process are very cool. Effectively, the excess wave breaking brings the flow of the upper vortex down to a speed at which atmospheric waves can no longer propagate, forming a 'critical surface'. This critical surface then sees additional wave breaking, which lowers the surface in altitude. This process repeats until the whole vortex has been destroyed, and waves can no longer propagate in the stratosphere.

to combine to their absolute maximum to produce enough wave forcing to disrupt the vortex.

So, what happened in the UK in 2018 and Chicago in 2019? Particularly strong wave activity in the northern hemisphere caused the stratospheric polar vortex to break apart in an SSW, for sure. But this was a change in the atmosphere ten kilometres or more from the surface. And while the polar vortex is huge, it doesn't have much weight to throw around. In order to impact on surface weather, it needs to collaborate with its tropospheric counterpart: the jet stream.

In 1999 it was discovered that when the northern hemisphere polar vortex breaks apart in an SSW event, the jet stream is diverted. Like a puppet with its strings loosened, in weeks following the SSW the jet stream drifts further south and becomes wavier. This stratospheric influence takes some time to reach the surface, typically around two weeks, but persists at the surface for two months on average. This is illustrated in the famous 'dripping paint' plot by Mark Baldwin and Tim Dunkerton.[10] Baldwin (or, as I knew him, Mark) was my PhD supervisor, and at every single conference I saw his plot in almost every presentation. It became something of a running joke in the stratospheric dynamics community that the dripping paint plot was inescapable. In their 1999 paper, Baldwin and Dunkerton identified that storms in the Atlantic tracked further south in the aftermath of an SSW. Subsequent work by the team also identified that cold weather became more extreme in the mid-latitudes, growing more likely and more intense. Both of these phenomena were a result of the wayward jet stream. The areas of low pressure formed by waves in the jet stream diverted storms at lower latitudes, bringing strong winds and intense precipitation in areas that otherwise wouldn't experience them. Equally, these meanders in the jet stream allowed frigid Arctic air to spill further south. This is exactly what happened in 2018 and 2019 – the British Isles and the Great Lakes were caught under a southward meander of the jet stream and froze under the freshly unchained Arctic air. Especially when combined with a source of moisture, such as an Atlantic gale, this is a perfect storm for wintertime chaos.

Previously we met Gilbert 'Boomerang' Walker, who meticulously calculated statistics for various fields in the atmosphere. In particular he was interested in establishing correlations between atmospheric fields in different locations, and identified several 'strategic points' of world weather. The Southern Oscillation in the Pacific was one such strategic point, influencing the Indian monsoon and weather patterns all over the world. Another strategic point was located in the North Atlantic, where he found another see-saw of atmospheric pressure, this time roughly between Iceland and the Azores. This came to be known as the North Atlantic Oscillation or NAO, and just like the Southern Oscillation it could also be described with a numerical index.[11] This NAO index broadly represents the influence of the jet stream on the weather in Europe. When the NAO index is high, the jet stream tracks further north than usual, and Europe experiences more fair weather, while if the NAO index is low, the jet stream tracks further south, and Europe experiences colder, stormier conditions.

To provide a little more detail, the work by Baldwin and Dunkerton found that there was a lagged connection between the strength of the stratospheric polar vortex and the value of the NAO index. There was some debate, however, about what exactly that meant. Could it be that the polar vortex was actually influencing the jet stream (via the NAO index), or was the more massive troposphere actually in the driver's seat, with the lag some kind of statistical artefact? To briefly insert myself into the story here, my PhD research was largely focused on this question of stratosphere–troposphere coupling, and attempting to unpick what was influencing what. It's still an area of research, but my thesis found that it was actually more complicated than simply saying one layer of the atmosphere influenced the other. In the aftermath of SSW events, I found that the stratosphere influenced the troposphere, and vice versa, but there was also a non-linear interaction between the two. Simply put, one layer told the other what to do, yes, but the conversation that happened between the two was just as – if not more – important in influencing weather.

What sets the Beast from the East and the Great Lakes freeze apart from, say, Hurricane Elena is that the associated extreme weather conditions with the former were well predicted. Authorities were given significant warning of the coming cold, and infrastructure was prepared for the extreme conditions. The reason this was possible was not because of superior weather prediction models or better-quality data, but because of the connection identified between the polar vortex and the jet stream. With the influence of an SSW typically taking a few weeks to propagate to the surface, meteorologists can observe an SSW taking place in the stratosphere and be on the lookout for tell-tale signs of trouble in the troposphere in the weeks ahead. Furthermore, computer models that used stratospheric information were found to have greater predictive skill than those using just tropospheric information.[12] In fact, some models see a greater improvement in forecast accuracy by including the stratosphere than by improving their spatial resolution. While tropospheric information – the location and intensity of storms, or surface temperature variations – does not persist for more than a few days at most, stratospheric information is rather more durable. The stratosphere varies on much longer timescales than the troposphere, and does so in ways that can be predicted much more accurately on long timescales. Hence, if you know what the state of the stratosphere is at a given time, you can be pretty confident what the state will be quite a long way into the future. In the month or two after SSWs occur, forecasters have extra information in their toolkits that can allow them to make weather predictions further into the future than they are otherwise able to.

This ability is not unique to the stratospheric polar vortex, but applies to other phenomena in the atmosphere, such as ENSO, as well as many others we've not discussed,* influencing weather at a great distance, and sometimes with a larger lead time. These processes are called *teleconnections*, and are a crucial part of modern meteorological predictions. That said, I hope you'll forgive me for the indulgence of this chapter, focusing specifically on my personal favourite subject in the atmosphere, the polar vortex. I hope that, if nothing else, it's introduced you to a new way of thinking about what influences the weather.

---

* A non-exhaustive list would include the QBO, the SAO, the MJO, and possibly – the consensus is trending in the opposite direction – the AMO. You can say a lot about atmospheric science, but you certainly can't say we lack acronyms.

CHAPTER 9

CHANGE

This book began with a certain young scientist camping in the woods of Big Sur, California. Finally, in this last chapter, we return to his story. But first, a little backstory.

At the tail-end of the eleventh century, Chinese polymath Shen Kuo (沈括) (1031–95) made a subtle observation about the Earth's climate. A landslide on the riverbank near modern-day Yan'an, Shaanxi province, revealed a cave, inside which was a large number of petrified bamboos. This was unusual, as bamboo does not, and did not, grow in northern China. Shen was intrigued, and came to the conclusion that while bamboo did not grow in the province at that time, at some point in the distant past it must have done. That meant, he deduced, that the climate in Shaanxi must have been significantly different during the past and, thus, that regional climates could change over time.[1] Shen was arguably the first person to write about what we now call *paleoclimate* – the climate of Earth in the distant past.

On a fundamental level, paleoclimate is something of a paradox. This is because, as we've seen, scientists draw a distinction between *weather* and *climate*. Weather is the variation of atmospheric conditions – rain, sun, cloudy, fog – as experienced on short timescales. This means what the atmosphere does day to day, week to week. Climate, by contrast, is the long-term average of atmospheric conditions, and hence the long-term average of weather. In the British Isles for example, we have many days that are cloudy and rainy, with temperatures between 10 and 20 °C. There are some days that are sunny and cloudless with temperatures above 20 °C, but when averaged over a long time, the weather in the UK is usually cloudy and mild. We therefore say that the British Isles has a cloudy and mild climate – this does not mean it doesn't have those warm, sunny days, but rather that they are far less common than the cloudy and mild ones.

The paradox is this: if climate is the long-term average, how can it change? How long is long term? How long do you need to average weather to obtain a climate? Officially, scientists use a period of thirty years to define a climate, as defined by the World Meteorological Organization, though in the more general sense of the word, climate is taken to mean how the atmosphere behaves on an infinitely long timescale. Yet, as Shen deduced nearly a millennium ago, this isn't actually the case. The climate at a given location, and indeed the climate averaged over the entire planet, can change quite dramatically.

This is a momentous idea! For this entire book we have examined how massive the atmosphere is, and how it contains a multitude of complex phenomena and interactions with other systems such as the oceans. If you had described much of the information in the previous chapters to a person from Ancient Greece or the Ottoman Empire, they would have had not too much trouble accepting it (though the part about digital computers may have been rather difficult to explain). Yet to explain to such a person the idea that the underlying climate, the very atmospheric giant itself, could change would be preposterous. The air around us could not change! It may be cloudy one day and sunny the next, but it will always still be the same sky, the same as it was for our forebears and will be for our children.

The story of the Earth changing on a grand scale is, of course, a story primarily told in *geology* (from the Ancient Greek for 'study of the Earth'). Much like atmospheric science, geology has deep roots, first written about by Ancient Greeks such as Aristotle and his successor Theophrastus (c.387–271 BCE) and later developed by Persian and Arab authors such as Ibn-Sina (نبا انیس) (981–1037) and Abu Rayhan al-Biruni (973–1050). Being the intellectual butterfly that he was, Shen Kuo also got in on the action, developing a theory of how land formed from the erosion of mountains and the deposition of silt by rivers. Meanwhile, in Europe, geology suffered from religious interference. It was believed that the Great Flood, the deluge that floated Noah and his ark, had formed the Earth as it was currently seen, and that the landscape did not change after this. As in many other areas of science, the powerful Catholic Church stifled innovation, prescribing truth via the Bible. The inventor of the barometer, Torricelli, suffered from the ministrations of the Church more than most, having his revolutionary mathematics of indivisibles suppressed.[2] It took until the mid-eighteenth century for geology to step out from the shadow of Christian doctrine, with influential texts arguing that natural processes shaped the planet, which must itself be much older than the Bible indicated.

At the start of the nineteenth century, it was becoming obvious that the Earth's climate must have significantly changed in the past. In particular, peculiar boulders in the French and Swiss Alps caught the attention of several geologists, who theorised that these *erratics* – boulders made of different rock to their surrounding geology – had been carried by the action of glaciers. Glaciers are, in essence, titanic rivers of ice, weighing as much as several hundred billion tons. While they appear to be solid, their extreme weight causes them to flow downhill very slowly, in much the same way as a chocolate bar left in your back pocket may look solid but is actually gooey and viscous and ruins your jeans. Scottish gentleman scholar James Hutton (1726–97), referred to by many as the father of modern geology,[3] proposed that erratics were caused by glacial retreat – the boulders had, in the past, been enclosed by ice, which then flowed downhill, carrying the rocks away from their native geologies. The 'mouth' of the glacier – the *ablation zone* – experiences loss of ice due to melting, calving, and other processes, and eventually the boulders would tumble out of the glacier when the surrounding ice melted.

However, many erratics were found at great distances from the nearest glacier. It is unclear if Hutton had an explanation for this,[4] but subsequent authors proposed a radical one: glaciers had previously covered a much greater part of Europe, and the climate was at this time much, much colder. The Swiss-born Louis Agassiz (1807–73) theorised that the Earth had previously experienced *Die Eiszeit* or an Ice Age, a cold period in the planet's history during which Alpine glaciers extended to cover much of Europe. In fact, Agassiz claimed, almost the entire northern hemisphere, including every northern continent, had been covered with ice sheets.[5] Whether Agassiz deserves all the credit for this idea is highly dubious. He didn't come up with the term himself – that was done by botanist and then-friend, Karl Friedrich Schimper (1803–67) – and his ideas also seem to have been pilfered from Schimper, who had been reluctant to publish them in the scientific literature but discussed them with Agassiz. Schimper was subsequently written out of Agassiz's accounts, the two having fallen out.[6] Clearly, sometimes even gentleman scientists behave more like the characters of *Mean Girls* than the austere gentlemen we see in paintings.

Regardless of who conceived it, the idea of an Ice Age was revolutionary. Agassiz went against the scientific consensus that the Earth was still cooling from the deep past when the planet was blisteringly hot, having just coalesced from accretion of material left over from the birth of the Sun.* How, the consensus asked, could the Earth cool and then warm again on such a large scale? Where was the energy coming from to allow such a dramatic swing in temperatures?

Several theories were in contention. Some scientists argued that perhaps the eruptions of volcanos changed the properties of the atmosphere and caused it to retain more heat. An alternative theory was presented by a remarkable Scotsman named James Croll (1821–90). Croll had one of the most unusual paths to a scientific career, and a similar story to that of his contemporary William Ferrel. Born on a farm in central Scotland and largely self-educated, he worked at various times as a wheelwright, a tea merchant, a hotel manager, and an insurance agent. His health constantly plagued him, and he seemed unable to hold down a job for long. Eventually, at the age of forty, he found himself working as a janitor at what is now the University of Strathclyde in Glasgow (then the Andersonian University).[7] As part of his duties, Croll had access to the university's extensive library. Stepping into the library for the first time must have been a fairy-tale moment for the middle-aged Croll, who, though sickly and coming from an underprivileged background, was fiercely intelligent and curious about the natural world. He was given permission to study using the books in the library, and quickly devoured their contents.

It really was like a fairy tale – the middle-aged hotelier-turned-janitor developed wide-ranging interests in philosophy, science, and theology, but of particular interest to him was the new

---

* Considering the hellish conditions on the planet at the time, the informal name for this period among geologists is the *Hadean*.

concept of ice ages. He found the ideas of Joseph Adhémar (1797–1862), focusing on the role of the Earth's orbit, intriguing, and within a few years he was publishing papers on the subject himself that drew the attention of the scientific establishment.

Every year the Earth completes a round trip about our star. In textbooks you will see this depicted as a circle around the Sun, but this is not quite correct – much as the Earth itself is not perfectly spherical, its orbit around the Sun is not circular but *elliptical*. An ellipse is a squashed, elongated circle, and handily we can define how squashed the ellipse of a planet's orbit is using a single number called the *orbital eccentricity*. A planet with an eccentricity of 0 has a perfectly circular orbit (in our solar system, the planet with the most circular orbit is Venus, with an eccentricity of just 0.007), while a planet with an eccentricity of 1 has a parabolic orbit, the bare minimum to just leave its solar system.

The Earth currently has an almost circular orbit, with an eccentricity of just 0.017. Croll argued that while this eccentricity was small, it could be significant enough to alter the Earth's climate. When the Earth is slightly further away from the Sun in its orbit, it receives less solar radiation, and so less heating. Equally, when its orbit brings it closer to the Sun, the planet receives more solar radiation and so more heating. In itself this should not produce a significant change in the Earth's climate – after all, the two different contributions should cancel out over the course of a year. However, the times of year that the orbit is close and far from the Sun slowly vary on an astronomical cycle – tens of thousands of years long – known as *apsidal precession*. Croll argued that there was an important feedback loop taking place. If the winter of a hemisphere coincided with the section of the orbit where the Earth was far from the Sun, the resulting extreme cold would cause extensive snowfall and ice formation. This would blanket the surface of the winter hemisphere in white, making the surface more reflective.

Scientists describe this as increasing the *albedo* (from the Latin for *white*) of the surface. This more reflective surface would then absorb less solar radiation, leading to deeper cold, and even more ice formation. If the resulting cold was extreme enough, ice could persist into the summer months, leading to year-round wintery conditions: an ice age. Eventually, the precession of the Earth's orbit would lead to warmer conditions in winter, and the ice age would give way to a warmer period. Croll's *ice albedo feedback* theory thus predicted alternating periods in the Earth's history – ice ages with ice-covered poles followed by interglacials with ice-free summers. Evidence was just starting to emerge from geologists that in fact the planet had indeed been through multiple cold periods, as Croll predicted. Unfortunately, however, Croll's predicted timings for these ice ages did not align with the geological evidence, fragmentary as it was at the time. It would take a further scientist, the brilliant Serbian/Croatian Milutin Milanković (1879–1958) to add the necessary extra complexity to Croll's theories. Milanković crucially corrected a mistake in Croll's argument - winters coinciding with a hemisphere being furthest in the Earth's orbit from the Sun was not the cause of the ice ages. Instead, it was when summers coincided with this furthest point in the orbit, reducing how much snow and ice thawed before the following winter. This reduced melt combined with the same quantity of snowfall in winter caused ice sheets to gradually accumulate. With this, and by accounting for additional elements in the complicated gravitational interactions between planets, accurately predicting changes in the Earth's eccentricity, axial tilt, and precession, using thousands of painstaking pencil-and-paper calculations, Milanković was able to explain observed ice ages in the historical record.

One component of this historical record is the collection of ice cores drilled in the Antarctic that we met earlier in this book. Other components include similar cores drilled in the ocean floor, the remains of small beetles (the wonderfully named field of palaeoentomology), and the growth rings in tree trunks. When these are combined, bringing data together from across the globe, it becomes very clear that the Earth has indeed experienced several periods of extreme cold, with much of the land being covered by ice sheets. However, there was a lot more going on than just the changes predicted by Milanković and Croll. The changes in solar radiation they predicted were present in the data, extremely clearly in the ice cores. Today we refer to these changes as Milanković cycles, in my opinion unfairly forgetting the remarkable James Croll.

These Milanković cycles are short-term changes when looking at the entire history of Earth. Over the past few hundred million years, the Earth has variously been a hothouse and a snowball, changing its climate on long, geological timescales and experiencing immense swings in global temperature of 20 °C or more. These swings could not have been caused by the orbital cycles derived by Croll and expanded on by Milanković, which vary on a scale of tens of thousands of years, not tens or hundreds of millions. Previously, we have discussed how energy from the Sun determines the temperature of the Earth. Could variations in the Sun's power output be responsible for the Earth's paleoclimate? However, while the power output of the Sun varies on both short and long timescales, neither match the variations in temperature observed in fossil records. In fact, the Sun has been very slowly increasing its power output over the past few billion years, while the Earth has largely been cooling during this time (with substantial variations).[8]

The reason for the Earth's paleoclimate varying so greatly in temperature is a puzzle that stretches back several hundred years. The solution began with a rather bizarre scientist obsessed with heat: the Frenchman Jean-Baptiste Joseph Fourier (1768–1830).

Fourier had a remarkable life. Orphaned at the age of nine, he originally trained for the priesthood before becoming a teacher, but he was imprisoned and close to being guillotined in the French Revolution.[9] Fortunately for science he was spared, and took up a professorship at the École Polytechnique in Paris.[10] For a time, he was then the governor of Lower Egypt after Napoleon's invasion of the country, and for much of the rest of his life he oscillated between political and scientific posts.

Reading Fourier's biography, you get the impression that he would have been quite happy just pursuing his research, but Napoleon, recognising his talents, kept dragging him back to civic appointments. During one such appointment in Grenoble, Fourier began experimenting with the propagation of heat. This was informed by his time in Egypt, where Fourier had decided that heat had life-giving properties (remember this, it becomes relevant). Using some innovative maths, Fourier was able to describe how temperature in an object evolves both in time and in space. The maths he developed, which now bears his name, had immense implications for the rest of science, but so too did the understanding of heat that his methods afforded. Of particular relevance to our story, Fourier, in the last few years of his life, concerned himself with the line of thinking we have been pursuing in this chapter, and found something disturbing: the Earth should be a lot colder!

We can quite easily calculate how warm the Earth should be. To do this we treat the Earth as a single thermodynamic object, and assume that it radiates energy into space in a way described by the equation for blackbody radiation that we met in Chapter 4. We then set this amount of energy to be equal to the total energy from the Sun absorbed by the Earth, which we know from observations. Working backwards from the blackbody radiation equation allows us to calculate the average temperature of the Earth if the only processes going on are absorption of solar radiation, and emission of thermal radiation by the Earth's surface. Doing this, we find a global average temperature of around *-18 °C*. Compare this to the global average temperature we observe, which is around 15 °C. The planet is *over 30 °C warmer* than basic thermodynamics says it should be! Fourier was troubled by this, and put forward several theories. He suggested, among other ideas, that perhaps 'interstellar radiation' supplied extra heat to the Earth. The theory that he eventually settled on, however, was that the atmosphere acted like an insulator, somehow providing the extra heating necessary to keep the Earth at habitable temperatures.

He never quite got to the bottom of this explanation, however: in a case of supreme irony, Fourier was killed by his obsession with heat. As an old man he kept his house overheated, and even wore excessively warm clothes to try to maximise the imagined healing power of heat. On 4 May 1830, possibly because he was faint after spending a day as a walking sauna, or simply due to tripping on his sweat-soaked dressing gown, Fourier fell down the stairs. He was badly injured, and died just a few days later.[11]

Most textbooks attribute the completion of Fourier's argument to have been completed in 1859 by Irish physicist John Tyndall (1820–1893),[12] and thanks to the latter's prestigious position at the Royal Institute in London he was undoubtedly important in mainstream acceptance of the hypothesis. However, this does a disservice to earlier laboratory work done across the Atlantic that demonstrated the important insulating properties of the atmosphere.

As described previously, light can be defined by its wavelength, and different substances will absorb different wavelengths. Ozone, for example, will absorb UV radiation, while oxygen and nitrogen will not. We can describe substances as being opaque or transparent to certain wavelengths of light – normally, of course, we'd use these words to reference whether an object absorbs visible light. A brick is opaque, while water is transparent, but only to visible light. However, we could just as validly discuss whether these objects are opaque or transparent to, say, ultraviolet light. Ozone, for example, is opaque to UV radiation, while nitrogen is transparent. At the time of Fourier's death, most scientists believed that all gases were transparent to infrared radiation, which has slightly longer wavelengths than visible light and is of crucial importance to most thermal radiation. However, in 1856, at the Eighth Annual Meeting of the American Association for the Advancement of Science in Albany, a two-page paper was presented by Professor Joseph Henry (1797–1878). This paper was years ahead of its time, detailing laboratory experiments that found certain gases were in fact opaque to infrared radiation, and that absorption of this radiation by these gases significantly increased their temperature. The gas that was found to produce the greatest warming was carbonic acid, better known by its modern name, carbon dioxide.[13]

This paper was not written by Henry, however – he was only presenting it. The experimental work had been conducted, and the analysis written up, by Eunice Newton Foote (1819–88). It would be fair to say that the scientific community of the nineteenth century, both in America and in Europe, was dominated by men. As we have already noted, this was largely thanks to the institutionalisation of science, with the formation of formal, governmental meteorological offices, as well as research universities and other institutions. The Smithsonian Institution, of which Joseph Henry was a professor, was founded 'for the increase and diffusion of knowledge among men',[14] and Henry himself felt obliged to append to his presentation of Foote's paper the note: 'Science was of no country and of no sex. The sphere of woman embraces not only the beautiful and the useful, but the true.'

While doubtlessly meant to praise Foote, this note only serves to highlight the exclusion of women from science at the time. This is not to say that no women participated in the scientific process – a huge number of celebrated female scientists from the Enlightenment onwards did so, such as Émilie du Châtelet, Caroline Herschel, and Mary Somerville. In fact, at the time of the Enlightenment women were much more active participants in science than in subsequent centuries.[15] By the nineteenth century, women carried out essential if undeniably menial tasks in science, such as mathematical calculations and cataloguing of observations. However, it was only in the mid-twentieth century that they were permitted to hold permanent positions in most leading research laboratories. By this time, the achievements of female scientists – even under extremely asymmetrical circumstances – had proved equal to and in many cases surpassing those of male scientists. Many know the names and accomplishments of great scientists from this era like Marie Skłodowska-Curie, Emmy Noether, Lise Meitner, and Ada Lovelace. While perhaps not as technically impressive as these greats, Eunice Foote should be just as well known. She was the first person on Earth to demonstrate that the atmosphere has the crucial property of absorbing infrared radiation. John Tyndall generally receives the acclaim for this discovery, and he did indeed independently come to the same result. But Foote beat him to the punch.

Fourier may not have had the experimental confirmation provided by Foote and Tyndall, but his hypothesis was correct: Earth's atmosphere acts as an insulator. While it is transparent to most of the light radiated by the Sun (mostly short wavelengths, the Sun being such a hot object), the atmosphere is a very effective absorber of the radiation given off by the Earth's surface (mostly long wavelengths, it being a cooler object). This absorption, as shown by Foote, is done partly by carbon dioxide, but mostly by water vapour. At any given time, the atmosphere contains over a thousand billion tons of water, both visibly in clouds and invisibly as water vapour.[16] This water is a very effective absorber of longer wavelengths of radiation, and contributes the dominant part of the atmosphere's insulating properties. Energy enters the Earth system from the Sun, mostly passing through the atmosphere, is re-emitted as thermal radiation by the Earth's surface and then prevented from leaving by water in the atmosphere, along with other trace compounds such as carbon dioxide and methane. This acts to keep the Earth much warmer than it would otherwise be, the disparity that Fourier initially calculated. This effect is what is commonly known as the *greenhouse effect*.

This term is often only used in reference to the *average* impact of the atmosphere, but one of the most significant effects of the atmosphere's insulating properties is only apparent after the Sun goes down. The night side of the Earth doesn't receive any energy from the Sun (which is blocked from view by the planet's rotation) yet continues to radiate energy into space. When the Sun goes down, this energy imbalance causes the temperature to drop significantly. In areas such as deserts, the variation between day and night can be as much as 30 °C (in certain conditions, even more – in a 'perfect storm' of weather conditions the hamlet of Loma in Montana experienced a change of temperature of 56.7 °C in just twenty-four hours.[17]) Generally, however, the Earth's surface experiences a day-to-night (or diurnal) variation in temperature of around 10 °C. This may seem quite significant, but thanks to the atmosphere it is actually quite minimal. If the Earth were stripped of its atmosphere, this diurnal variation in temperature would be many, many times greater. Much the same line of reasoning can be applied to annual variations in temperature too: the difference between summer and winter is much less with the presence of an atmosphere than on a planet without one.

For what it's worth, the term greenhouse effect was actually coined much later than Fourier's initial work, by John Henry Poynting (1852–1914) and Frank Very (1852–1927) in a series of feuding papers published in 1909 and 1910. It's a rather unfortunate term, as it's not terribly accurate –indeed, this imprecision lay behind the disputes between Poynting and Very carried out in scientific papers – and highlights the need for clear communication by scientists. In 1901 the Swedish meteorologist Nils Ekholm (1848–1923) published a paper that explained the insulating properties of the atmosphere.[18] In this he had used the term 'greenhouse' to make an analogy: the atmosphere prevented heat loss into space in much the same way that the glass of a greenhouse does, except that the glass of a greenhouse prevents heat loss by keeping air trapped in one place, stopping convection transporting warm air – and so thermal energy – away. The atmosphere, on the other hand, prevents energy leaving the planet in the form of thermal *radiation*. The end result is the same, but the mechanism is rather different.

Poynting wrote a paper arguing about this analogy, using the term 'greenhouse effect' in quotation marks, though unfortunately making a bit of a mess of the calculations involved in the actual thermal transfer. Very then published a response in the same journal, using 'greenhouse effect' in the title of his paper. Between the two of them, with a special credit to Ekholm, the world was introduced to the catchy – if flawed – term to describe how the atmosphere insulates the Earth. Note that this is an entirely natural phenomenon, however. Much present discussion of it is about the man-made (anthropogenic) greenhouse effect, which we will come to much later, but the insulating effect Ekholm, Poynting, and Very were arguing about is the entirely natural greenhouse effect caused by carbon dioxide and water vapour already in the atmosphere.

The amount of carbon dioxide in our atmosphere has varied greatly over the last billion years or so. Presently, around 400 parts per million (ppm) of the air we breathe is carbon dioxide, meaning that if you equally divided a volume of air into a million equal parts, 400 of those parts would be entirely carbon dioxide. For reference, around 781,000 of those million parts would be nitrogen, and 201,000 would be oxygen. Carbon dioxide is really quite a small part of the atmosphere, but it has an immensely oversized impact. It should also be stated that, in the past, the concentration of carbon dioxide was much higher. Scientists estimate that during the Cambrian Period, about 500 million years ago, the CO_2 concentration might have been as great as 4,000 ppm. This decreased to a minimum in the Permian Period, around 300 million years ago, increased over the following few hundred million years, around the time of the dinosaurs, and then decreased again to the modest current value.

These changes in carbon dioxide concentration correlate with changes in the Earth's average temperature described in Chapter 2. This follows from Fourier's argument – the more insulating material you put in the atmosphere, the greater the amount of heat that is trapped, and so the higher the temperature. Scientists who have studied Antarctic ice cores have been able to measure the CO_2 concentrations over the past several hundred thousand years extremely accurately, and compare these measurements with those of global average temperatures over the same time period.

So far, so good. However, the reverse is also true: when the temperature increases, so too does the concentration of CO_2. This muddies the waters somewhat. This has led some people to suggest that changes in CO_2 lag behind changes in temperature, which is only partially correct. In the case of the past several hundred thousand years, the observed changes in temperature have been largely due to changes in the Earth's orbit – Milanković cycles. When the planet comes out of an ice age, the amount of energy it receives from the Sun increases, melting ice and warming the oceans. The process of doing this causes the oceans to release CO_2 into the atmosphere, which then goes on to cause *additional* warming.

The ocean releases CO_2 when warmed for chemical reasons – the warmer water is, the less CO_2 can be stored in it. You can test this for yourself by taking two cans of carbonated drink, keeping one in a fridge and the other at room temperature. After a few hours, open both the cans. The can in the fridge will hiss when opened due to the CO_2 previously kept bound up in the drink escaping. The can that was kept at room temperature will hiss much more loudly however, as the water in the drink will have been unable to retain all the CO_2 it originally contained, and have outgassed it to the air pocket inside the can. When the can is opened, this extra gas rushes out, causing a louder hiss.

It's estimated that about 90 per cent of the total warming experienced when Earth comes out of an ice age occurs after the increase in CO_2 given off by the oceans,[19] so while it is true that some CO_2 increase does lag behind increases in temperature, the vast majority of warming follows an increase in atmospheric carbon dioxide. In the recent past, this relationship has been muddied by variations in the Earth's orbit dipping the planet into and out of ice ages. In the more distant past, when the planet was much warmer and hence not experiencing regular glacial/inter-glacial periods, the principal driver of climate was the varying concentration of CO_2 in the atmosphere. This has somewhat kicked the question can down the road yet again, however. We have explained why the Earth's average temperature has varied so much over its history: due to changes in carbon dioxide. But why did atmospheric CO_2 vary so much during the Earth's history? We're talking about truly enormous changes, so you would expect the cause to be similarly epic. The truth is actually rather banal. If you're reading this in the UK, the answer probably lies outside your window: rain.

When rain falls through the atmosphere, it picks up very small amounts of carbon dioxide to create a weak carbonic acid. This removes carbon from the atmosphere, and when the rain runs off into the oceans, it puts the carbon into deep storage. This storage might be as carbonic acid in the ocean, or in the Earth's interior as it gets pulled down into the mantle at destructive tectonic plate boundaries. Alternatively, if the rain falls on volcanic rock, it is absorbed straight into the land, while if it lands on carbonate rock it actually slightly dissolves the surface and releases more carbon dioxide into the atmosphere. Eventually, carbon that is in deep storage will be re-emitted into the atmosphere via volcanic activity and constructive tectonic plate boundaries.

In short, there is a geological carbon cycle that takes place over millions and millions of years. Depending on the configuration of land and sea – whether the continents lie close to the equator or towards the poles, or if areas with carbonate rocks coincide with areas of high precipitation – the balance of carbon in the atmosphere and in the deep storage of the oceans and crust can alter over long timescales. This internal variability, with all the turning speed and inevitability of an oil tanker, has ultimately caused the concentration of atmospheric CO_2 to change, leading to monumental shifts in the Earth's climate over the past billion years or so.

At this point, we have a pretty good picture of the Earth's past climate. We know not only that it has changed significantly, but also why. On long, geological timescales, these changes are due to the global carbon cycle, and then modified and amplified by other factors such as James Croll's ice feedback cycle, and the chemistry of carbon in water. On shorter timescales of tens or hundreds of thousands of years, these changes are due to variations in the Earth's orbit that are spectacularly amplified by processes internal to the planet. We just so happen to live in a time of relatively little carbon dioxide in the atmosphere, and hence these orbital changes can have enormous effects.

Yet, there is one extra timescale to consider. When examining the last few hundred years of CO_2 data, scientists noticed something puzzling: the concentration of carbon dioxide in the atmosphere was increasing. About 250 years ago, the amount of CO_2 trapped in the air bubbles was around 280 ppm. Over time this concentration increased slowly at first, but the rate of change grew faster and faster. By the present day the concentration of CO_2 had increased to over 400 ppm. This observation has been confirmed by other techniques, including modern-day observations directly measuring the CO_2 content of air samples. The trend was unmistakable – there was more and more carbon dioxide in the atmosphere every year. Yet this increase did not match any known natural cycle. It could not be explained by the glacial pace of orbital change, and certainly not by the geological pace of carbon cycling.

Whatever could be happening?

Theophrastus' *On the Causes of Plants* might seem an unlikely place to find the first hints of the end of the world. Written at the turn of the fourth century BCE, and mostly concerned with types of trees, shrubs, and cereals, Theophrastus notes that in several places in the eastern Mediterranean where trees have been felled or marshes drained, the local weather was consistently different compared to before these alterations.[20] The air was colder, frost more frequent, and the local wine ruined. In other words, the climate local to these areas had been altered by the economic activity of humans. Several other writers through history made similar observations, such as extraordinary Prussian polymath Alexander von Humboldt, founder of Humboldtian science that so influenced men like Glaisher. While travelling in South America, von Humboldt noted that deforestation of parts of the Amazon rainforest in Venezuela was causing a dramatic fall in the level of Lake Valencia.[21] He later wrote:

> When forests are destroyed, as they are everywhere in America by the European planters, with an imprudent precipitation, the springs are entirely dried up, or become less abundant, the beds of the rivers remaining dry during a part of the year, are converted into torrents, whenever great rains fall on the heights.

Von Humboldt was looking at the relationship between humans and the natural world like no one ever had before. The knowledge that technology had advanced to the point where humans could alter the fabric of the planet was enthusiastically seized upon.

As the nineteenth century steamed on, roads and railways were constructed, rivers stoppered by dams, and the landscape rebuilt to meet our needs. In the most extreme displays of this power, oceans were connected by engineering brute force at Panama and Suez – waters that had been separated for millions of years could now flow. Science and technology could accomplish anything! Humans were seen as masters of nature, not merely stewards, enabled by their supreme technology. Man (and it was always man) had the power to create and destroy as he saw fit, and the Earth would simply have to adapt.

This power had erupted out of the workshops of Europe in the eighteenth and nineteenth centuries. What had enabled this dramatic change? In a word, *coal*. Coal is, in essence, ancient sunlight trapped in rock. Its origins lie in the Carboniferous Period, about 360 to 300 million years ago, when an order of plants known as the *Lepidodendrales* (Ancient Greek for 'scale tree') came to dominate the skylines of Earth. These were some of the earliest tree-like plants, some reaching fifty metres in height, possessing large leaves and thick trunks, but relative shallow roots. This meant that these proto-trees fell over with comic frequency. One of the most remarkable things about these plants, however, was that their wood was made of a certain type of polymer called *lignin*. This structural material, along with cellulose, gave plants the ability to stand tall and rigid, and led to the explosive spread of trees over dry land.

However, as plants evolved the ability to produce lignin, a crucial part of the ecosystem took a while to catch up with this development: bacteria.[22] When *Lepidodendrales* were falling in Carboniferous forests, there were no bacteria waiting on the forest floor to break down the lignin. Instead, the fallen wood just sat there, untouched, for millennia. Over time, more and more trees would fall on top of one another and compress the earlier trees, first into peat and eventually into coal. The sunlight that these plants had drunk in over their lifetimes, converted to carbon-based molecules by photosynthesis, was then trapped in this rock. The huge quantities of carbon buried in the ground during this time is from where the Carboniferous Period takes its name – and while coal formation continued for hundreds of millions of years after, it's estimated that some 90 per cent of the coal we use for fuel today comes from these 'coal forests'.[23]

While it was still locked in the ground, this huge deposit of carbon had no impact on the rest of the world. For 300 million years, it formed part of the immense stores of carbon in the Earth's crust. Eventually, however, a species of animal evolved that valued the buried carbon highly enough to rip it from the ground.

The year 1776 was marked by revolution. Of minor significance was a colony declaring independence from its European masters (though we will come back to this). Far more significant was the revolution in the British Isles. As was so frequently the case, the modern world was being ushered in by a Scotsman,* and, in this case, that Scotsman was James Watt (1736–1819). Watt came from a privileged family, working in his father's shipwright yards for a time before setting up a workshop in the University of Glasgow, repairing and maintaining scientific instruments. Frequently, this involved delicate astronomical instruments or lab equipment.

One day, however, Watt was asked to repair a small model of a steam engine. Perhaps this is surprising to you! Despite popular perception, Watt did not actually invent the steam engine: the first steam engines that modern eyes would recognise were developed earlier, in the seventeenth century, though most were theoretical apparatuses rather than practical ones. The earliest recorded industrial use for a steam engine is in Spain, where Jerónimo de Ayanz y Beaumont (1533–1613), a member of the nobility, built and used a rudimentary steam engine to drain a silver mine near Seville in 1611.[24] It is unclear how successful this was, though the fact that the design was not subsequently used elsewhere indicates

---

* I'm not kidding, Scotland largely created the modern world. A non-exhaustive list of things invented by Scots includes: television, telephones, antibiotics, bicycles, electrodynamics, steamships, tarmac, steam hammers, the *Encyclopaedia Britannica*, logarithms, golf, ice hockey, the refrigerator, colour photography, the flushing toilet, vacuum flasks, geology, pneumatic tyres, reciprocating steam engines, and, of course, Irn-Bru.

that perhaps it was not terribly so. Sadly, Ayanz died only a few years afterwards, and did not have a chance to improve his potentially revolutionary idea.*

The first practical steam engine was instead invented by Thomas Newcomen (1664–1729) an English ironmonger. His invention – the Newcomen engine – was the spark of the Industrial Revolution. His was a huge device that generated steam in a sealed cylinder using a coal-fired boiler. Cold water was then injected into the cylinder, condensing the steam and dramatically decreasing the pressure in the cylinder. The cylinder head, motivated by this partial vacuum from below and atmospheric pressure from above (hence the name *atmospheric engine*), descended to compress the air in the cylinder. In so doing, the downward motion of the engine generated mechanical force that was used to raise the other end of a beam balanced on a fulcrum. Eventually, the weight of the beam would raise the cylinder head, and steam would be allowed to enter into the cylinder again. This cycle was then repeated around twelve times per minute, producing the power equivalent to twenty horses.[25]

Newcomen's engine was a success, and was used across England to pump water out of mines. By simply burning a lump of black earth, without the labour of men or horses, mines could be kept dry and productive. Those mines could then be used to produce more metal and more coal, making more money and, crucially,

---

* As historian Anton Howes notes, what we know about Ayanz's design indicates that it was extremely inefficient to the point of complete impracticability. Ayanz was simply too early to understand how air pressure worked correctly, and so, quoting Howes, 'Ayanz may have invented a steam engine, but he did not invent the atmospheric engine'. See A. Howes, 'Age of Invention: The Spanish Engine', 24 July 2020. [Online: available at https://antonhowes.substack.com/p/age-of-invention-the-spanish-engine (accessed 9 November 2020).]

making more steam engines possible. This could catch on. There was a problem, however – the single, jerky power stroke of the Newcomen engine made it extremely difficult to power machinery, and so its use was limited to applications such as pumping water. The engines also guzzled coal thanks to their simplistic design. The fires of the Industrial Revolution had not yet been lit, but the sparks were flying.

When presented by the University of Glasgow with a small model of Newcomen's engine to repair, Watt realised that he could improve its design. After extensive tinkering, he added a separate chamber where water could condense, keeping it at a constant low pressure. This, and a few other technical additions, spectacularly improved the efficiency of the engine.

With the condensation taking place separately from the cylinder, Watt was also able to make the engine *push* as well as *pull*. Combining this with his parallel motion (a mechanical linkage that translated up-and-down motion into rotating motion), Watt allowed the steam engine to drive rotary machinery. This would prove to be an earth-shattering accomplishment. Watt's steam engine, first commercially installed in 1776, at an ironworks near Falkirk, Scotland, could now drive steamships, factories, pumps, and railway engines. British industry now had the ability to convert coal into raw economic productivity. With ample coal supplies available within the country, Great Britain led the charge into the Industrial Revolution. The friendly, genius engineer who by his own admission 'would rather face a loaded cannon than settle an account or make a bargain'[26] had fundamentally remade mankind's relationship with nature. Humans were coming for the natural world, now powered by steam.

Before long, other countries started using Watt's steam engine. As the nineteenth century progressed, improvements were made to its design, allowing for engines to run on higher steam pressures, with greater power and efficiency. The world ran on steam, and hence, on coal. The air in European cities was thick with coal smoke and other emissions from industry. In 1873 in London some 700 people died from the greenish-yellow polluted air, at its worst resembling pea soup.[27] These local changes were obviously attributable to industry, but as this was the source of the wealth of empires, it was an acceptable price to pay. The air quality was so poor, and the emissions of factories so extreme, however, that some scientists began to wonder just how much coal was being burned by factories around the world. Perhaps humans were souring the air of the entire planet? It seemed a ludicrous idea, the atmosphere surely being so large as to be indifferent to human affairs. The nature of science is to enquire, however, and some scientists took to the task.

Swedish geologist Arvid Högbom (1857–1940) compiled estimates of how carbon cycled through natural processes, such as that emitted by volcanic activity, taken up by oceans, or released by acidic rain, among others. It occurred to him in 1896 to also include man-made carbon emissions, from factories and railways and so forth. After extensive calculations he was surprised at the result. Human activities were adding roughly as much CO_2 to the atmosphere as natural processes already were.[28] These man-made, or anthropogenic, carbon emissions were still miniscule compared to the amount of CO_2 already in the atmosphere, however. Reassuringly, he estimated that the total carbon emitted from burning coal in 1896 would raise the concentration in the atmosphere by only a thousandth part. But if the emissions continued long enough, or if they increased, then perhaps this could be significant.

At this time, Högbom was in conversation with another Swedish academic, the walrus-moustachioed chemist Svante

Arrhenius (1859–1927). After a distinguished career in chemistry, Arrhenius found himself interested in Earth's ice ages. He wondered if he could derive a theory using physical chemistry to explain how the Earth periodically experienced freezing conditions. In particular, he was interested in the role of CO_2. He hypothesised that if for some reason the concentration of carbon dioxide in the atmosphere increased – say, after a large volcanic eruption – because of the heat-trapping effect of the molecule, global temperatures would slightly increase as a result.

This slight increase would lead to a far more important consequence: the warmer air would hold more moisture. The additional water vapour in the air would then enhance the warming, which could cause more vapour to be taken up into the atmosphere, causing more warming, and so on. Conversely, if the level of CO_2 were to decrease, a small amount of cooling would take place, and the air would hold less water vapour. This would cause more significant cooling and, if this feedback loop continued, potentially tip the Earth into an ice age.

This was the atmospheric chemistry equivalent of James Croll's ice albedo feedback loop from a few decades prior, indicating that even small changes in CO_2 concentrations could produce cascading effects in global temperature. All it took was a spark to start a roaring fire. Such complex effects were far beyond Arrhenius's ability to calculate in full, but he committed himself to trying. Simplifying the world down to a few latitude bands, and using data on the absorption of infrared radiation by gases that was primitive by today's standards, Arrhenius spent months arduously calculating using pencil and paper, possibly as a deliberate distraction to his ongoing divorce.

Eventually, he published his results. If the concentration of CO_2 in the atmosphere – a very minor gas by volume, remember – were to halve, the overall effect on global temperatures would be a cooling of around 5 °C.[29] Conversely, if the concentration of atmospheric CO_2 doubled, the Earth would warm by perhaps 5 or 6 °C. Informed by Högbom's work, Arrhenius predicted that if humanity emitted greater amounts of carbon into the atmosphere in the future, the global average temperature would gradually increase. He saw this as a great relief, however, as this could counteract any new ice age that was looming on the horizon:

> Is it probable that we shall in the coming geological ages be visited by a new ice period that will drive us from our temperate countries into the hotter climates of Africa? There does not appear to be much ground for such an apprehension. The enormous combustion of coal by our industrial establishments suffices to increase the percentage of carbon dioxide in the air to a perceptible degree ... By the influence of the increasing percentage of carbonic acid in the atmosphere, we may hope to enjoy ages with more equable and better climates, especially as regards the colder regions of the earth, ages when the earth will bring forth much more abundant crops than at present, for the benefit of rapidly propagating mankind.[30]

Other scientists were sceptical about Arrhenius's predictions. They pointed out that he had over-simplified the Earth, ignoring myriad complexities, including the effects of clouds in response to changing levels of water vapour. More clouds would surely reflect more light away from the Earth, balancing out the additional insulation provided by water vapour. Additionally, they claimed, it was impossible for CO_2 to build up in the atmosphere. If humanity added any of the stuff, it would simply be absorbed into the vast carbon storage of the ocean, removed from the air. Our impact was simply too miniscule compared to natural processes. Nature (with a capital N) was eternal, immortal and separate to Man (with a capital M). How could man-made soot possibly affect divine creation? By the turn of the century, most scientists had dismissed Arrhenius as being entirely wrong.

Over the course of this book, something that I hope has become clear is that science is often not truly the product of just individuals. While individuals may make remarkable accomplishments, propelling our understanding of the natural world forward, these accomplishments are enabled only by circumstance. The wealth of a nation, the availability of particular materials, the quality of education given to the general populace – broader societal factors position and enable individuals to make their contributions. We have seen with James Croll and William Ferrel, in particular, the effect of scientific textbooks and journals being made widely available in the nineteenth century. Any history of science is truthfully a history of these societal factors. Sometimes though, there are figures who make a contribution almost despite of their circumstances, rather than because of them. Without these extraordinary men and women and their persistence, our wealth of knowledge would be poorer. One of these men, one of the most remarkable people in the entire story of the atmosphere, and returning to our revolutionary colony, was Charles David Keeling (1928–2005). We have, of course, already met him.

Born in Pennsylvania, Keeling studied chemistry, earning a PhD in polymerisation from Northwestern University in 1953. The first societal pull he resisted was that of the booming petrochemical industry. Rather than continue into a lucrative job, as most of his PhD cohort did, Keeling instead stayed in academia, moving to the geochemistry department of the California Institute of Technology (Caltech). Geochemistry is the field that studies the chemical processes underlying the Earth, such as in rock formation, and how elements such as carbon cycle through a planet's crust, oceans, and atmosphere.

Keeling enjoyed hiking through the mountains of the American northwest, and this passion for the outdoors continued through his entire life. On discovering geochemistry, a way to combine his chemical knowledge with this desire to be in the open air, he was clearly smitten. Originally, his research at Caltech focused on whether carbonate in rivers and ground water was in equilibrium with the CO_2 in the air, but soon he turned his attention to accurately measuring that atmospheric CO_2 concentration.[31]

Keeling was good at this. He designed his own equipment, allowing him to measure CO_2 concentrations more accurately than anyone else, and used this to justify travelling across the United States to collect air samples. His first trip was to Big Sur State Park in California. Here among the redwoods, by a gentle stream, he began the lifelong activity of unstoppering glass flasks and taking a small sample of air that could later be processed in the lab. Our story of the atmosphere begins and ends, it seems, with glass.

On his trips Keeling took air samples every few hours, and from these repeated measurements he identified a number of interesting things. First, carbon dioxide concentrations varied throughout a day – they were lower during the daytime when plants sucked in the gas for photosynthesis, and higher at night when those same plants respired and produced it. Second, carbon dioxide was remarkably consistent. From Maryland to California, CO_2 was always between 315 and 320 parts per million of the air in his samples. By, frankly, larking around the continental United States, Keeling had found something significant – CO_2 was well mixed in the atmosphere away from man-made sources such as factories and freeways. This indicated that there was a single value of average carbon dioxide concentrations in the atmosphere. To get an accurate measurement of it, Keeling simply needed to be as far away from human activity as possible.

To this end, in March 1958 Keeling began daily measurements of CO_2 at an observatory on the north slopes of Mauna Loa, Hawaii. This was using new equipment, however, and originally he thought there was a problem with it. His measurements fluctuated wildly over the course of the year, peaking in May at around 315 ppm and falling to a low of 310 ppm in November. When the concentration began to rise again in December, Keeling realised that he had in fact discovered another cycle of carbon in the atmosphere. Oxygen concentrations in the atmosphere rise and fall with the activity of photosynthesis, and so CO_2 concentrations rise and fall in opposition to this. As there is far more landmass, and so far more plants, in the northern hemisphere, this meant CO_2 peaked in the northern hemisphere's summer. Keeling was in effect measuring the planet breathing, photosynthesis dominating in the northern hemisphere's summer, only to be replaced by respiration in the winter. Each year should have been easy to predict, a regular sine wave of CO_2 concentration across the months. This wasn't the case, however.

When Keeling published his first data in 1960, it seemed the second year of measurements indicated a slightly higher concentration than the first year. The sine wave pattern didn't quite join up – the concentration in May one year was not the same as the concentration in May the next. This was no big deal, Keeling thought; perhaps there was some other cycle at play that simply needed more time to become evident. So, he kept taking measurements. The third year of observations produced a still higher concentration of CO_2. Then the fourth year did the same.

Something was clearly up: the concentration of carbon dioxide in the atmosphere was measurably increasing year on year.

Keeling was originally funded by money left over from the International Geophysical Year (1957–58), an ambitious programme of cooperative science projects across the world, from the Antarctic to the outer reaches of the atmosphere. This money did not last forever, and within a few years administrators wanted Keeling to call time on his observation programme. The second societal pull he resisted, then, was the Cold War. As far as the government was concerned, an accurate measurement of global carbon dioxide concentrations had now been made, and it was time to start funding other projects. At this time, the United States government was spending considerable money on geoscience research, mainly to gain an edge over the Soviet Union. The politics of the time were drawing money towards astronautics, seismology, weather prediction, seemingly everywhere other than measuring trace gases in the atmosphere (that, as far as the government was concerned, had already been measured quite accurately enough). Keeling was having none of it, convinced that these measurements were important and must be continued. For the next four decades he constantly fought to keep the observatory open, scrounging money from different agencies and officials. Except for a hiatus from February to April 1964, the Mauna Loa observatory has kept continuous measurements from 1958 right up to the present day.[32]

The 'Keeling Curve', as these measurements have become known (see Figure 11), show as clear as day the annual cycle in carbon dioxide concentrations: a sine wave wiggling with a period of exactly one year. On top of this variability, however, utterly dominating the graph, is the steady increase in global carbon dioxide concentration. Prior to the invention of the steam engine, the concentration was estimated to be 280 parts per million (ppm). In 1958, Keeling measured the concentration to be 315 parts per million. By the time of his death – of a heart attack while hiking on his Montana ranch in 2005 – that measurement was 377 parts per million, almost 20 per cent more than when he started.

Keeling's single-minded determination, against all societal

forces and perhaps against many peoples' idea of time well spent, gifted science one of the most significant datasets of all time. While society wanted to move on, first pulling Keeling towards the oil industry, and later away from his research towards more militaristic applications, he persisted.

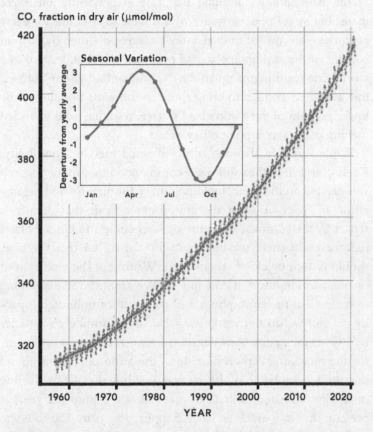

MONTHLY MEAN CO$_2$ CONCENTRATION

Figure 11: *The Keeling Curve of carbon dioxide concentrations at Mauna Loa, Hawaii.*

But Keeling had not discovered global warming. Rather, he had discovered the *possibility* of global warming. Building on the work of Fourier, Foote, Tyndall, and Arrhenius, he had showed that the concentration of carbon dioxide in the atmosphere was increasing, and that, as per Arrhenius's calculations, this *could* alter global climate. In 1963 he and a few other scientists published a report ominously titled 'Implications of Rising Carbon Dioxide Content of the Atmosphere', detailing that carbon concentrations were increasing by 0.7 ppm per year.[33] If this were to continue for a few centuries – less time if the rate were to increase – then the carbon content of the atmosphere would eventually double, leading to a worldwide warming of up to 3.8 °C. According to the authors, this would be 'enough to bring about an immense flooding of the lower portions of the world's land surface, resulting from increased melting of glaciers' among other effects.

While certainly alarming, this remained merely a possibility. Keeling and the other authors recommended that more research was needed on the topic, including more funding for and organisation of observation of the atmosphere. Later that decade, a report by the National Academy of Sciences in the United States indicated that there was no cause for alarm, but that the issue should indeed be closely monitored. Writing of the emissions of carbon from industry, it concluded: 'We are just now beginning to realise that the atmosphere is not a dump of unlimited capacity . . . but we do not yet know what the atmosphere's capacity is.'[34] Over the next several decades, scientists continued to monitor the increasing carbon dioxide in the atmosphere, and found that the accumulation of the gas was speeding up. While Keeling originally estimated the rate of increase as 0.6 ppm per year, at present this rate is closer to 2.5 ppm per year. Carbon was undeniably building up. But was there any evidence that the Earth was responding as Arrhenius and others predicted?

Simple physics indicated that an increase of a gas such as CO_2 in the atmosphere would trap more heat, and raise global temperatures, but this was still an overly simplistic argument. As Arrhenius's critics correctly pointed out, there are additional factors controlling global temperature. We have already discussed some of these – including the ice albedo feedback articulated by James Croll, and the water vapour feedback set out by Arrhenius himself. In the second half of the twentieth century, a baffling array of additional, more complex feedbacks were discovered by scientists. Aside from the heavy hitters, such as warming leading to more water vapour leading to more clouds leading to less sunlight reaching the ground leading to cooling (i.e. cloud albedo feedback), there are other, more subtle interactions between carbon concentrations and global temperatures to consider. To give just one example, scientists have recently found that when levels of carbon dioxide in the atmosphere rise, plants respond by slightly thickening their leaves.[35] While we don't yet know why they do this, we do know that in so doing they become less efficient at sucking carbon out of the atmosphere. Higher carbon levels thus lead to a less efficient removal of carbon from the atmosphere. So, when atmospheric carbon levels are elevated for whatever reason, there is an additional positive feedback – in addition to water vapour, ice albedo, and so on – pushing carbon levels even higher.

This is all to say that in the mid-twentieth century there was still a huge uncertainty in how Earth would respond to any change in carbon dioxide levels. Would there be significant global warming, or even global cooling in response to changes in CO_2? The proof would be in the data. Fortunately, by the mid-twentieth century, there was no lack of it. Thousands of meteorological stations around the world had decades of temperature records, though they were all taking measurements at different times, using different units, and with different methodologies. Several research groups attempted to untangle this data spaghetti, removing unreliable sources and standardising the results of disparate weather stations. This was, needless to say, a monumental task. The first group, headed up by James Hansen (1941–), showed that by 1980 the world had warmed by 0.2 °C compared to pre-industrial temperatures. This was a small signal, nearly swamped by year-to-year variability. There was also a small cooling trend from 1940 to 1960,* which was quite clear in the data. Hansen's group was soon joined in this conclusion by report after report from other research groups, and with each passing scientific paper published, the evidence became more and more concrete.

To organise and collate all this research, the World Meteorological Organization and the United Nations environmental agencies created the Intergovernmental Panel on Climate Change (IPCC) in 1988. A common misconception of the IPCC is that it puts out reports written by politicians to further a political agenda of some kind. This is not true. IPCC reports – and we've had six major ones to date – are written by

---

* This gave rise to the myth that scientists flipped from warning of global cooling and an impending ice age in the 1970s to warning of global warming and the end of the world in the 1980s. This is *partly* true – for example, even Hansen's group identified that the northern hemisphere had experienced widespread cooling in the mid-century, bucking the trend of overall warming. Meanwhile, the southern hemisphere had consistently warmed. Further, an analysis of scientific literature from the 1950s to the present shows that there wasn't a single year in which the consensus – represented by research papers published – suggested global cooling lay in the future. There was certainly healthy debate in the middle of the century as to whether climate sensitivity to CO_2 was positive or negative, but within a few decades it was clear that the overall effect of all different climate feedbacks was global warming. See T. Peterson, W. Connolley and J. Fleck, 'The Myth of the 1970s Global Cooling Scientific Consensus', *Bulletin of the American Meteorological Society*, vol. 89, no. 9 (2008), pp. 1325–38.

scientists, who, incidentally, are unpaid for this work, and these reports are then subject to line-by-line scrutiny by representatives of national governments. In other words, the IPCC was, and is, an interesting hybrid political and scientific organisation, condensing research conducted by scientists into periodic reports curated by politicians. Thanks in particular to representatives from fossil-fuel-exporting (and -dependent) nations, this has the result of diluting the conclusions of the scientists, and tends to lead towards very conservative conclusions,[36] but even with this restriction, by the turn of the twentieth century, the scientific consensus was clear: the world was warming, and in such a way that could only be explained by man-made emissions of carbon dioxide.

The full story of how this was accomplished is fascinating, with many colourful characters and twists and turns, but this book is long enough as it is, so for the full story I refer you to Spencer Weart's *The Discovery of Global Warming*.[37] Suffice to say, by the late twentieth century there was ample evidence that the planet had already warmed in response to man-made carbon dioxide. Quite why society was not immediately galvanised into action, limiting further emissions and thus further warming, instead of prevaricating about the reliability of the science, is a story best told in Oreskes and Conway's *Merchants of Doubt*.[38] To say it again for the people at the back, for several decades now the science has been very, very clear. This is not a recent conclusion. It is not a hasty conclusion. Scientists have been saying this for decades, with plentiful evidence. Humans have appreciably raised the concentration of atmospheric carbon dioxide, and this has directly caused changes in our climate.

To be more specific, at the time of writing, the concentration of CO_2 in the atmosphere is 414 ppm, an increase of nearly 50 per cent relative to the pre-industrial concentration of 280 ppm.

Accompanying this change in CO_2 concentration has been a warming across the globe of, on average, approximately 1 °C.[39] Besides temperature, however, there have been a whole raft of other changes to the global climate – to rainfall patterns, to the frequency of extreme events, the average sea level, the severity of storms, and so on. These changes are referred to as *climate change*, while the simpler metric of global average temperature increasing is referred to as *global warming*. I would argue that the difference between these two – very similar – terms is this: global warming is abstract; climate change is experienced.

One of many reasons that humanity has struggled to accept that we are influencing the Earth's climate is that the changes to date have been so slow, and so small. Summer being one degree Celsius warmer than it was for our great-great-great-great grandparents hardly sounds like a serious issue. It is not *experienced*. This is only the case when looking on a global scale, however. Examine a particular locality and the story is very different. To pick just one example, consider the wildfire season in the western United States. Since the mid-twentieth century, this season has grown in length by two and a half months. Of the ten years with the most wildfire activity on record, nine have occurred since the year 2000.[40] This is not a small, slow change. To people living in the western United States, changes in wildfires have absolutely affected their lived experience in the past few decades. This is also just one example – climate change has impacted on how people live their lives via drought, famine, conflict, storms, and any number of natural disasters. Or perhaps I should say 'natural' disasters?

The effects of humans' influence on the climate are not evenly distributed, with those hardest hit so far perversely the ones who have caused the fewest carbon emissions. Those in the industrialised world have largely been shielded from the consequences of their actions and those of their ancestors, though, as in the instance of US wildfires, this is not entirely the case any more. In ways that science is still struggling to comprehend, the atmosphere appears to be responding to our carbon emissions. In some places this brings more rain, in others much less. In many places it raises the sea level, flooding coastal areas and forcing people inland. Species of animals are migrating en masse as their previous ranges become uninhabitable, or disappearing altogether. Diseases are finding new localities to spread to. Facing scarcity, humans find themselves forced to migrate. International tensions over fresh water and arable land rise and rise.

The atmospheric giant is facing a challenge unlike any other in its multibillion-year history. For so long it has varied on geological timescales, free to adapt to changes over tens of thousands of years. Individual cells and organs in the giant might shift dramatically, even violently on small scales, but its fundamental physiology has remained constant. Since the invention of the steam engine, however, it has felt its feet pressed to the fire. It has never, ever faced such a rapid change in its conditions. Year after year the fire burns hotter and hotter, and so the giant makes increasingly defiant attempts to wriggle free. By the present day its patience has run out. What will happen next is difficult to predict. But it will not be pretty.

Everything thus far in this book has been fact. As all science must be, it is supported by data, published in peer-reviewed journals, verified, and correct according to the most rigorous process we have at our disposal. The IPCC reports in particular are the most rigorously reviewed and meticulously researched documents in the history of academia. As far as we can say *anything* in science is true, the findings of the IPCC on climate change to date are.

That is the story to date, however.

In this final section I want to briefly discuss the future of the atmospheric giant. Necessarily, this discussion will be filled with probabilistic statements such as 'likely' or 'almost certainly'. I have no doubt that many of these predictions will turn out to be incorrect, or even that the information necessary to make such predictions will be out of date by the time you read this paragraph. If you happen to be reading this in a future century, I certainly hope that you look back on these predictions with wry humour as by-products of an overly pessimistic, hand-wringing group of scientists. Yet I have a sneaking suspicion that you may in fact be looking back on them as having underestimated the risks, and underplayed the severity of the situation that had developed since the invention of the steam engine.

The question I'm most frequently asked by friends on the topic of climate change is, 'Just how bad is it?', closely followed by the related, 'Is it hopeless?', as if the giant was a sick patient and we its worried family in the waiting room. To be clear, the patient is fine. The Earth has experienced much higher carbon levels in the past, and has survived far more severe heat than we could possibly inflict upon it. The atmospheric giant will survive this, and will certainly survive us. That's not the issue here. The last time carbon concentrations were as high in the atmosphere as they are now was during the Pliocene, some three million years ago. During this time, global temperatures were 3 or 4 °C warmer, and sea levels twenty metres higher than they currently are.[41] A sea level rise of twenty metres would be, to use technical language, 'game over'. At present, approximately half a billion people live within twenty metres of the sea, and while these people would obviously need to move, so too would those affected by tidal surges even further inland. Farmers whose fields would flood too frequently to make agriculture sustainable would cease business, reducing food production and adding to migratory pressures.

Society would be placed under impossible stress to continue as we know it.

A comparison with the Pliocene offers some idea of where we are headed in the future. Given enough time, the atmosphere – at current CO_2 concentrations – will equilibrate at a temperature similar to those in the Pliocene. This assumes, however, that we have emitted as much carbon into the atmosphere as we are ever going to, which is hardly a good assumption. It is likely that global carbon emissions will peak in the mid-twenty-first century, though they may plateau rather than decline. This means that by the end of the century the concentration of CO_2 could be as high as 800 ppm or, in a best-case scenario, as low as 425 ppm. As with all predictions of climate change, it is incredibly difficult to give an accurate figure – emissions are particularly difficult to estimate as they are not governed by deterministic laws of physics but by the decidedly fuzzy and human-driven laws of economics. This, when combined with the huge uncertainties that still remain about how the world will respond to an unprecedented shock in atmospheric CO_2 concentrations, means that all predictions made by scientists have large error bars.

What is likely to happen by 2100? A summary that scientists can be reasonably confident in is this: carbon dioxide concentrations will likely peak between 500 and 600 ppm. This will lead to a global warming of around 2 °C.[42] Changes that are expected to take place as a result of this include: average sea levels rising by two metres, category five hurricanes doubling in frequency, several billion people losing ready access to fresh water, tropical diseases becoming far more common further from the equator, an increase in migration and resource-based conflicts, and a mass extinction of plant and animal species occurring on a scale that has not been seen while humans have existed on Earth. It is almost certain that Earth will look very different, and be much poorer for it, in 2100. The list I have just described is also by no means exhaustive, and it certainly shouldn't be interpreted as sensationalist or cherry-picking the upper end of predictions. If anything, that list is a conservative reading of the literature – many scientists predict a far, far more grim, dark future.* To be clear though, I mean a grim, dark future for *us*. The atmosphere and the planet, if not the life currently clinging to its surface, will live on and survive. At present, humanity is sawing off the branch it is sat on – through its current activities, in particular relying on fossil fuels such as coal and now oil, it is souring the environment that it is reliant upon to live. We are, as a species, pushing the global climate further and further away from the climate that we evolved to cope with, and that we need to survive.

This is not to say that things are completely hopeless! There is a great deal we can do to reduce our emissions of carbon into the atmosphere, and so reduce the future impacts of climate change. More than this, we can use this opportunity to value more of what remains of the natural world, and even to restore what has previously been lost to industry. The principal component of this will be a transition away from fossil fuels towards renewable technologies, not just for electricity generation but also for transport and for heating and cooling our homes. While the common image of this process is a wind turbine or a solar panel, it should really be an electric heat pump or electric boiler. Fully half of all energy we use as individuals is ultimately used for heating and cooling, and only 10 per cent of this energy currently comes from renewable sources.[43] A societal transition towards more renewable energy generation is already in progress

---

* All this and much more is compiled in David Wallace-Wells's soberingly titled *The Uninhabitable Earth: A Story of the Future*, London: Penguin, 2019.

and accelerating, but there is a great deal more that can be done, and indeed must be done, to avoid the worst possible outcomes of climate change. Contrary to what many believe, combatting climate change does not necessitate a global state policing what we can and cannot do, or the kinds of light bulbs we can and cannot use. Greater legislation supporting renewable energy, steering us away from fossil fuel use, and requiring greater energy efficiency, is certainly necessary. But this does not need to descend from an aloof central global government such as the United Nations. This problem requires *cooperation*, not *unity*. And it will certainly be tough. But with enough effort, we are in a position to curb our emissions and limit the damage we are inflicting upon ourselves. As a society we can start to value the natural environment, value the giant, as essential to our very survival, and turn back from the path we set on with the invention of the steam engine.

Ultimately, we are at the mercy of the giant. It is not our enemy any more than a collection of bacteria might consider us their enemy. The atmosphere is entirely indifferent to our existence. Perhaps when it sits down to write its multibillion-year autobiography, we will feature as a minor footnote, a blip lasting a few thousand years characterised by a massive spike in carbon dioxide followed by a long, slow, whimpering decline. But maybe even this is overstating our importance.

We need it. It does not need us.

FAMILY

One of the most remarkable pictures I have ever seen is held at the British Library in London. It was engraved by artist Ferdinand Helfreich Fritsch (1707–58) for *Theoria Motuum Planetarum et Cometarum*, a book written by Leonhard Euler (1707–83) explaining the motions of the planets and comets through a hypothetical, all-pervasive fluid. The frontispiece of the book depicts the solar system as it was then known: six planets with associated moons orbiting a brilliant Sun. A comet orbits on a roguish path, inclined steeply to the orbital plane. What makes this image so extraordinary is what lies beyond the solar system: Fritsch depicts a huge number of other stars, also with their own solar systems. Planets orbiting other suns are illuminated by starlight, tracing out their orbits. As two angels unfurl the fabric of the universe, we see this scene playing out across the cosmos, alien planets orbiting alien suns seen and unseen, obeying the same mechanical rules established for our very own solar system, here on Earth.

Published in 1744, this engraving predated the first discovery of a planet around another star – an *exoplanet* – by two and a half centuries.*

---

* The idea that there are worlds around other stars is actually even older, arguably emerging with Giordano Bruno (1548–1600) in the sixteenth century.

Figure 12: *F.H. Fritsch's engraving of exoplanetary solar systems from 1744.*

There was no way that Fritsch, or indeed Euler, could have predicted that so many alien worlds would exist in our stellar neighbourhood. At the time of writing, ninety-seven exoplanets have been discovered within just 10 parsecs (a mere 300 thousand billion kilometres: our backyard in galactic terms).[1] Some of these worlds are truly strange. Some are known as 'hot Jupiters' – gas giants that orbit so close to their stars that their atmospheres are superheated and vanishing out into the void. Others are more recognisable to us – gas giants like Saturn and Jupiter, and terrestrial planets similar to Earth and Mars. Many are tidally locked, orbiting their star at the same rate that they spin on their axes. These worlds possess one scorching hot hemisphere and another that is locked in endless night, the boundary between the two being a fascinating temperate environment. Perhaps, if conditions are right, such an environment would be habitable.

Throughout this book we have discovered what makes our atmosphere tick – the anatomy and physiology of our very own atmospheric giant. Once we only looked at its footsteps. Now, hopefully, we appreciate its complexities, understand its musculature, can see the entire giant in all its splendour.

The beautiful thing about physics, however, is that it is the same everywhere.

Our giant is just one of an extended family – it has ninety-seven relatives in the local area, and millions further afield. As described in this book, by understanding how energy flows through the atmosphere, how air masses move across the surface and are deflected, and how distinct layers form in the atmosphere, we have developed an immensely powerful toolkit that can be used on other worlds. The application of this knowledge has already begun. The decidedly underweight atmospheric giant of Mars has been monitored using satellites and simulated using computer models, and found to have many of the same features as our giant. Mars has polar vortices, governed by the same equations as on Earth. Titan, the moon of Saturn, has a frozen atmospheric giant made of nitrogen and methane, wreathed in clouds made of hydrocarbons. As alien as this world sounds, it too obeys the same equations derived by pioneers like William Ferrel and Joseph Fourier, kept much warmer by its own greenhouse effect.

We have very limited information about planets that orbit other stars, but thanks to some clever analysis of the light that passes through their atmospheres we can conclude a few things. Some of them are made of decidedly unusual materials. A super-sized Jupiter-like planet called HAT-P-7b was found to probably contain clouds of vaporised rubies and sapphires.[2] This is caused by the temperature on the planet's surface being around 2,000 °C, orbiting its star so closely that a year lasts only two Earth days. While their material is unusual, the behaviour of these clouds can be determined using the same fluid physics we apply to describe clouds on Earth.

In fact, using numerical models and the physics developed to model Earth's atmosphere, just with different parameters, we can take what we do know about these planets and hypothesise about how their atmospheres behave. For example, tidally locked planets have been simulated, and it is believed that, if water is present, strong convection would lead to extensive cloud formation, lessening the difference between the day and night sides.[3] Such planets are serious contenders for human habitation, and in fact due to the size and temperature of the most common types of stars, we expect *most* habitable Earth-like planets to be tidally locked. Should we ever visit these alien worlds, there are doubtlessly further amazing features to discover that we have never even considered trying to simulate. These planets would be home to unique combinations of parameters in the universal equations of fluid motion that we have never seen before, producing atmospheric circulations we can only dream of.

Our giant has a family. Really quite a large family, in fact. Thanks to thousands of experimentalists and theorists throughout history, without ever setting foot on the surfaces of alien worlds we can still understand how their giants behave. This is a truly amazing thing! Gazing up at the night sky, perhaps sat by a tent in Big Sur, California, we can pick a star and imagine a world around it, calculate its expected temperature, even its expected weather. Whether or not we ever use this information practically to colonise worlds in this solar system or others, we have already transcended the limitations of this planet. Perhaps our atmospheric giant will be only the first of many that we come to know very well – or perhaps not.

I hope that over the course of this book I have convinced you of a few things. First, that atmospheric science has a history as long and noble as any other field of research. Like many sciences it can trace its roots back to antiquity, though it is really in the Renaissance and the Early Modern period that it came into its own. It has strong connections with chemistry, physics, and geology, and some of the greatest scientists in history have contributed to our knowledge. Equally, there have been a number of figures who made crucial contributions that should be better known. People like the janitor-turned-academic James Croll, Eunice Foote, discoverer of the insulating properties of CO_2, and the wide-eyed farm-boy William Ferrel. Similarly, important developments were made away from traditional Western academia by individuals such as Shen Kuo and Wasaburo Ooishi. This is a global science, far more than just recent developments in numerical weather prediction and climate science in Western universities. In the tapestry of world knowledge, study of the atmosphere is a thread that weaves through the fabric, densely in some places, sparsely in others, of human history.

Second, atmospheric science did not develop smoothly. As with all science, it is based on data, and this data only became available in fits and bursts. Originally, researchers were limited to what they could perceive themselves with their senses and the written accounts of others, but the development of key technologies allowed for entirely new perspectives on the atmosphere. First, the invention of Venetian glass in the years leading up to the Renaissance – along with other technological advancements – allowed for enclosing small samples of the atmosphere, and the construction of specialist scientific instruments. Our giant could now be measured and quantised. This was only possible on small scales, however, limited by where a scientist could take measurements themselves. Then, the ascendancy of European colonial empires and the birth of proto-globalisation concentrated information gathered over large areas into centralised organisations. At first these were military or commercial, such as the East India Company, and later they included official government agencies, inspired by Humboldtian science. This process was considerably accelerated by the development of the telegraph (and later internet) and, by the twentieth century, data collection was advanced enough to allow for accurate predictions of what the atmosphere would do in the near future. New technologies such as gas-filled balloons and liquid-fuelled rockets then took our understanding further, allowing for a truly global perspective of weather, and the discovery of immense features in the atmosphere such as the polar vortices.

Perhaps there is no better example of these arguments in miniature than James Glaisher. Coming to the field of atmospheric science at a critical time in its by then already illustrious history, he embraced the use of technology. He did so in the manner of the age-old explorer, risking his own life to reach new heights in the atmosphere. He did so in an innovative way, collating atmospheric information over a wide area near-instantaneously. He paved the way for future balloonists to discover the upper layers of our atmosphere, and provided the data that enabled the first forecasting of our weather. Yet these individual accomplishments were only made possible by the development of technology, itself only made possible by the social and economic revolutions of the eighteenth and nineteenth centuries. His most lasting contribution, to the world of forecasting, was only enabled by the institutional structure in which he operated. Glaisher, the individual, accomplished remarkable things, but those accomplishments were, to borrow a phrase, bubbles on the tide of society.

Finally, I hope I have convinced you that atmospheric science is essential. The invention of another technology – the steam engine – has forced us to confront a new reality. Our atmosphere is changing, and we are responsible. Science has shown us that the planet has been really quite different in the deep past, and that with continued emissions of carbon dioxide the climate will shift away from the careful balance of conditions we evolved to survive. How we avoid devastating changes in our climate is a question not just of atmospheric science – it involves economics, sociology, politics, and many other fields – but is a question that ultimately begins and ends *with* atmospheric science. Without detailed knowledge of how the planet will respond to changes in atmospheric composition, it is impossible to plot a course of action out of our situation. Rather like Captain Robert FitzRoy, we find ourselves in treacherous waters, but we do have our instruments, and we know how to use them. Granted, there are still many unanswered questions, many uncertainties in our current knowledge, but we have a pretty good idea of what to expect if we do nothing. The miraculous, beautiful, complex atmosphere that currently sustains us could erase society as we know it in a matter of centuries, if not decades. That may sound like hyperbole, but unfortunately it is not. The science indicates our situation is grave, and growing worse with every passing year of insufficient action. Fortunately, we have centuries of accumulated knowledge at our disposal. We have thousands of passionate scientists still building on that knowledge. It is ultimately down to this generation, including you reading this book right now, to take our knowledge of the giant and act upon it.

We cannot assume that another giant will take us in, and shield us from the cosmos. Humans already have one, happy to keep us safe and warm, fed and watered. In return, we must now use the collected knowledge of the past 500 years to keep it on our side.

We've had our time to appreciate the beauty here. Now we must protect it.

ENDNOTES

Chapter 1: Idea

1 R. Holmes, *Falling Upwards: How We Took to the Air*, London: William Collins, 2013.

2 J. Glaisher, *Travels in the Air*, London: Bentley, 1871.

3 H. Zinszer, 'Meteorological Mileposts', *Scientific Monthly*, vol. 58, no. 4 (1944).

4 G. Wainwright, *The Sky Religion in Egypt*, Cambridge: Cambridge University Press, 1938.

5 R. Wilkinson, *The Complete Gods and Goddesses of Ancient Egypt*, London: Thames & Hudson, 2003.

6 G. Hellmann, 'The Dawn of Meteorology', *Quarterly Journal of the Royal Meteorology Society*, vol. 34, no. 148 (1908).

7 H. Frisinger, *The History of Metereology: To 1800*, New York: Science History Publications, 1977.

8 ibid.

9 H. Frisinger, 'Aristotle and his "Meteorologica"', *Bulletin of the American Meteorological Society*, vol. 53, no. 7 (1972), pp. 634–8.

10 A. Gregory, *Eureka! The Birth of Science*, London: Icon Books, 2001.

11 W. Napier Shaw, *Manual of Meteorology*, Cambridge: Cambridge University Press, 1926.

12 S. Rasmussen, 'Advances in 13th Century Glass Manufacturing and their Effect on Chemical Progress', *Bulletin for the History of Chemistry*, vol. 33, no. 1 (2008), pp. 28–34.

13 H.C. Bolton, *Evolution of the Thermometer 1592–1743*, Easton, PA: The Chemical Publishing Co., 1900.

14 C. Huygens, *Oeuvres completes de Christiaan Huygens publiees par la Societe Hollandaise des Sciences*, The Hague, 1893.

15 D. Fahrenheit, 'Experimenta et observationes de congelatione aquae in vacuo factae', *Philosophical Transactions of the Royal Society*, vol. 33 (1724), pp. 78–89.

16 A. Alexander, *Infinitesimal: How a Dangerous Mathematical Theory Shaped the Modern World*, London: Oneworld Publications, 2014.

17 J. West, 'Torricelli and the Ocean of Air: The First Measurement of Barometric Pressure', *Physiology*, vol. 28, no. 2 (2013), pp. 66–73.

Chapter 2: Birth

1 Y. Yan, M. Bender, E. Brook, H. Clifford, P. Kemeny, A. Kurbatov, S. Mackay, P. Mayewski, J. Ng, J. Sveringhaus and J. Higgins, 'Two-million-year-old Snapshots of Atmospheric Gases from Antarctic Ice', *Nature*, vol. 574 (2019), pp. 663–6.

2 K. Zahnle, L. Schaefer and B. Fegley, 'Earth's Earliest Atmospheres', *Cold Spring Harbor Perspectives in Biology*, vol. 2, no. 10 (2010).

3 A. Zerkle and S. Mikhail, 'The Geobiological Nitrogen Cycle: From Microbes to the Mantle', *Geobiology*, vol. 15, no. 3 (2017), pp. 343–52.

4 K. Tyrell, 'Oldest Fossils Ever Found Show Life on Earth Began Before 3.5 Billion Years Ago', University of Wisconsin-Madison, 18 December 2017. [Online: available at https://news.wisc.edu/oldest-fossils-found-show-life-began-before-3-5-billion-years-ago/ (accessed 2 November 2020).]

5 H. Holland, 'The Oxygenation of the Atmosphere and Oceans', *Philosophical Transactions of the Royal Society of London B: Biological Sciences*, vol. 361, no. 1470 (2006), pp. 903–15.

6 B. Schirrmeister, J. de Vos, A. Antonelli and H. Bagheri, 'Evolution of multicellularity coincided with increased diversification of cyano-bacteria and the Great Oxidation Event', *Proceedings of the National Academy of Sciences of the United States of America*, vol. 110, no. 5, (2013) pp. 1791–6.

7 R. Berner and Z. Kothavala, 'Geocarb III: A Revised Model of Atmospheric CO_2 over Phanerozoic Time', *American Journal of Science*, vol. 301, no. 2 (2001), pp. 182–204.

8 D. Royer, R. Berner, I. Montañez, N. Tabor and D. Beerling, 'CO$_2$ as a Primary Driver of Phanerozoic Climate', *GSA Today*, vol. 14, no. 3 (2004).

9 B. Goldstein, 'Ibn Muʿādh's Treatise on Twilight and the Height of the Atmosphere', *Archive for History of Exact Sciences*, vol. 17, no. 2 (1977), pp. 97–118.

10 JAXA, 'Research on Balloons to Float over 50 km Altitude', Institute of Space and Astronautical Science, 2008. [Online: available at http://www.isas.jaxa.jp/e/special/2003/yamagami/03.shtml (accessed 18 November 2020).]

11 M. Lehman, *Robert H. Goddard: Pioneer of Space Research*, New York: Da Capo Press, 1988.

12 W. von Braun, 'Recollections of Childhood: Early Experiences in Rocketry as Told by Werner Von Braun 1963'. [Online: available at https://web.archive.org/web/20090212140739/http://history.msfc.nasa.gov/vonbraun/recollect-childhood.html (accessed 12 July 2020).]

13 M. Neufeld, *Von Braun: Dreamer of Space, Engineer of War*, New York: A.A. Knopf, 2007.

14 S. Ramsey, *Tools of War: History of Weapons in Early Modern Times*, Delhi: Vij Books, 2016.

15 M.J. Neufeld, *The Rocket and the Reich: Peenemünde and the Coming of the Ballistic Missile Era*, New York: The Free Press, 1995.

16 N. Best, R. Havens and H. LaGow, 'Pressure and Temperature of the Atmosphere to 120 km', *Physical Review*, vol. 71, no. 12 (1947), pp. 915–16.

17 N. Best, R. Havens and H. LaGow, 'Pressure and Temperature of the Atmosphere to 120km', *Physical Review* (1947), pp. 915–16.

18 F.A. Lindemann and G.M.B. Dobson, 'A Theory of Meteors, and the Density and Temperature of the Outer Atmosphere to which it Leads', *Proceedings of the Royal Society of London. Series A, Containing Papers of a Mathematical and Physical Character*, vol. 102, no. 717 (1923), pp. 411–37.

19 F. Götz, A. Meetham and G. Dobson, 'The Vertical Distribution of Ozone in the Atmosphere', *Proceedings of the Royal Society of London. Series A, Containing Papers of a Mathematical and Physical Character*, vol. 145, no. 855 (1934), pp. 416–46.

Chapter 3: Wind

1 University of Warwick, '5400mph Winds Discovered Hurtling Around Planet Outside Solar System', 13 November 2015. [online: available at https://warwick.ac.uk/newsandevents/pressreleases/5400mph_winds_discovered/ (accessed 4 August 2021).]

2 D. Defoe, *The Storm*, London, 1704.

3 M. Walker, *History of the Meteorological Office*, Cambridge: Cambridge University Press, 2012.

4 'William C. Redfield 1789–1857', *Weatherwise*, vol. 22, no. 6 (1969), pp. 225–62.

5 C. Abbe, 'Memoir of William Ferrel: 1817–1891', Cambridge, 1892.

6 W. Ferrel, 'On the Effect of the Sun and Moon Upon the Rotatory Motion of the Earth', *Astronomical Journal*, vol. 3 (1853).

7 W. Ferrel, 'An Essay on the Winds and Currents of the Oceans', *Nashville Journal of Medicine and Surgery*, 1856.

8 W. Ferrel, 'The Influence of the Earth's Rotation Upon the Relative Motion of Bodies Near its Surface', *Astronomical Journal*, vol. 5, no. 109 (1858), pp. 97–100.

9 M. Buys-Ballot, 'Note sur le rapport de l'intensité et de la direction du vent avec les écarts simultanés du baromètre', *Académie des sciences (France) Comptes rendus hebdomadaires*, vol. 45 (1857), pp. 765–8.

Chapter 4: Fields

1 G. Vallis, *Atmospheric and Oceanic Fluid Dynamics*, Cambridge: Cambridge University Press, 2006.

2 S. Blundell and K. Blundell, *Concepts in Thermal Physics*, Oxford: Oxford University Press, 2010.

3 Y. Matsumi and M. Kawasaki, 'Photolysis of Atmospheric Ozone in the Ultraviolet Region', *Chemical Review*, vol. 103, no. 12 (2003), pp. 4767–82.

4 Global Ozone Research and Monitoring Project, 'Scientific Assessment of Ozone Depletion: 2018', WMO, 2018.

Chapter 5: Trade

1 G. Cawkwell, *Philip of Macedon*, London: Faber & Faber, 1978.

2 D. Sobel, *Longitude: The True Story of a Lone Genius Who Solved the Greatest Scientific Problem of His Time*, London: Harper Perennial, 2011.

3 T. Woollings, *Jet Stream: A Journey Through Our Changing Climate*, Oxford: Oxford University Press, 2019.

4 P. Frankopan, *The Silk Roads*, London: Bloomsbury, 2015.

5 A. Hopkins, *Globalization in World History*, New York: Norton, 2002.

6 G. Hadley, 'Concerning the Cause of the General Trade Winds', *Philosophical Transactions of the Royal Society of London*, vol. 39, no. 437 (1735), pp. 58–62.

7 T. Woollings, *Jet Stream: A Journey Through Our Changing Climate*, Oxford: Oxford University Press, 2019.

8 J. O'Connor and E. Robertson, 'Gaspard Gustave de Coriolis', MacTutor, July 2000. [Online: available at https://mathshistory.st-andrews.ac.uk/Biographies/Coriolis/ (accessed 5 October 2020).]

9 G. Coriolis, 'Sur les équations du mouvement relatif des systèmes de corps', *Journal de l'École Royale Polytechnique*, vol. 15 (1835), pp. 144–54.

Chapter 6: Distance

1 M. Kottek, J. Grieser, C. Beck, B. Rudolf and F. Rubel, 'World Map of the Köppen-Geiger Climate Classification Updated', *Meteorologische Zeitschrift*, vol. 15, no. 3 (2006), pp. 259–63.

2 Staff members of the Department of Meteorology, 'On the General Circulation of the Atmosphere in Middle Latitudes', *Bulletin of the American Meteorological Society*, vol. 28 (1947), pp. 255–80.

3 T. Woollings, *Jet Stream: A Journey Through Our Changing Climate*, Oxford: Oxford University Press, 2019.

4 H. Seilkopf, 'Maritime Meteorologie: Vol. 2', in *Handbuch der Fliegerwetterkunde*, Radetzke, 1939, p. 359.

5 J. Lewis, 'Ooishi's Observation Viewed in the Context of Jet Stream Discovery', *Bulletin of the American Meteorological Society*, vol. 84, no. 3 (2003), pp. 357–70.

6 J. Lewis, 'Ooishi's Observation Viewed in the Context of Jet Stream Discovery', *Bulletin of the American Meteorological Society*, vol. 84, no. 3 (2003), pp. 357–70.

7 T. Woollings, *Jet Stream: A Journey Through Our Changing Climate*, Oxford: Oxford University Press, 2019.

8 UK Met Office, 'What is Saharan Dust?' [Online: available at https://www.metoffice.gov.uk/weather/learn-about/weather/types-of-weather/wind/saharan-dust (accessed 20 July 2021).]

9 J.M. Wallace and P.V. Hobbs, *Atmospheric Science: An Introductory Survey*, Academic Press, 2006.

10 K. Singh, *I Shall Not Hear the Nightingale*, New York: Grove Press, 1959.

11 L. Ahman, R. Kanth, S. Parvaze and S. Mahdi, *Experimental Agrometeorology: A Practical Manual*, Cham, Switzerland: Springer, 2017.

12 W. Dalrymple, *The Anarchy: The Relentless Rise of the East India Company*, New York: Bloomsbury Publishing, 2019.

13 J. Hickel, 'How Britain Stole $45 Trillion from India', 19 December 2018. [Online: available at https://www.aljazeera.com/opinions/2018/12/19/how-britain-stole-45-trillion-from-india/ (accessed 18 October 2020).]

14 M. Davis, *Late Victorian Holocausts: El Niño Famines and the Making of the Third World*, London: Verso, 2000.

15 B. Fagan, *Floods, Famines, and Emperors*, London: Pimlico, 2000.

16 ibid.

17 J. Bjerknes, 'Atmospheric Teleconnections from the Equatorial Pacific', *Monthly Weather Review*, vol. 97, no. 3 (1969), pp. 163–72.

18 C. Wang, C. Deser, J.-Y. Yu, P. DiNezio and A. Clement, 'El Niño and Southern Oscillation (ENSO): A Review', in *Coral Reefs of the Eastern Pacific*, Springer Science, (2016) pp. 85–106.

19 C. Ropelewski and M. Halpert, 'Global and Regional Scale Precipitation Patterns Associated with the El Niño/Southern Oscillation', *Monthly Weather Review*, vol. 115 (1987), pp. 1606–26.

20 K. Kumar, B. Rajagopalan, M. Hoerling, G. Bates and M. Cane, 'Unraveling the Mystery of Indian Monsoon Failure During El Niño', *Science*, vol. 314, no. 5796 (2006), pp. 115–19.

21 B. Fagan, *Floods, Famines, and Emperors*, London: Pimlico, 2000.

Chapter 7: Forecast

1 NOAA, 'Hurricanes: Frequently Asked Questions', 1 June 2021. [Online: available at https://www.aoml.noaa.gov/hrd-faq/] (accessed 20 July 2021).

2 National Hurricane Center, 'Hurricane Elena Preliminary Report', National Oceanic and Atmospheric Administration, Miami, 1985.

3 T. Fort, *Under the Weather*, Century, 2006.

4 K. Teague and N. Gallicchio, *The Evolution of Meteorology: A Look in the Past, Present, and Future of Weather Forecasting*, Oxford: Wiley, 2017.

5 A. Wulf, *The Invention of Nature: Alexander von Humboldt's New World*, New York: Knopf, 2015.

6 P. Moore, *The Weather Experiment: The Pioneers Who Sought to See the Future*, London: Chatto & Windus, 2015.

7 M. Walker, *History of the Meteorological Office*, Cambridge: Cambridge University Press, 2012.

8 ibid.

9 ibid.

10 UK Parliament, 'The Witchcraft Act', 1735.

11 K. Teague and N. Gallicchio, *The Evolution of Meteorology: A Look in the Past, Present, and Future of Weather Forecasting*, Oxford: Wiley, 2017.

12 G. Vallis, *Atmospheric and Oceanic Fluid Dynamics*, Cambridge: Cambridge University Press, 2006.

13 R. Friedman, *Appropriating the Weather: Vilhelm Bjerknes and the Construction of a Modern Meteorology*, Ithaca, NY: Cornell University Press, 2018.

14 J. Fleming, *Inventing Atmospheric Science: Bjerknes, Rossby, Wexler, and the Foundations of Modern Meteorology*, Cambridge, MA: MIT Press, 2016.

15 J.M. Wallace and P.V. Hobbs, *Atmospheric Science: An Introductory Survey*, Academic Press, 2006.

16 UK Met Office, 'Global Accuracy at a Local Level'. [Online: available at https://www.metoffice.gov.uk/about-us/what/accuracy-and-trust/how-accurate-are-our-public-forecasts (accessed 20 July 2021).]

17 L. Richardson, *Weather Prediction by Numerical Process*, Cambridge: Cambridge University Press, 1922.

18 P. Edwards, *A Vast Machine: Computer Models, Climate Data, and the Politics of Global Warming*, Cambridge, MA: MIT Press, 2010.

19 S. Strogatz, *Nonlinear Dynamics and Chaos: With Applications to Physics,*

Biology, Chemistry, and Engineering, Boulder, CO: Westview Press, 2015.
20 E. Lorenz, *The Essence of Chaos*, London: UCL Press, 1993.
21 E. Lorenz, 'Deterministic Nonperiodic Flow', *Journal of the Atmospheric Sciences*, vol. 20, no. 2 (1963), pp. 130–41.
22 C. Danforth, 'Chaos in an Atmosphere Hanging on a Wall', 2013. [Online: available at http://mpe.dimacs.rutgers.edu/2013/03/17/ chaos-in-an-atmosphere-hanging-on-a-wall/ (accessed 7 October 2020).]
23 UK Met Office, 'Global Accuracy at a Local Level'. [Online: available at https://www.metoffice.gov.uk/about-us/what/accuracy-and-trust/how-accurate-are-our-public-forecasts (accessed 20 July 2021).]

Chapter 8: Vortex

1 AON, 'Global Catastrophe Recap', 2018.
2 CBS Chicago, 'It's Official, Chicago Is Colder than Parts of the Arctic, Yukon, and Mars', 30 January 2019. [Online: available at https://chicago.cbslocal.com/2019/01/30/chicago-deep-freeze-colder-than-arctic-yukon-mars-siberia-mount-everest/ (accessed 20 November 2020).]
3 F. Whipple, 'The Propagation of Sound to Great Distances', *Quarterly Journal of the Royal Meteorological Society*, vol. 61, no. 261 (1935).
4 C. Choi, 'Strange But True: Earth Is Not Round', *Scientific American*, 12 April 2007. [Online: available at https://www.scientificamerican.com/article/earth-is-not-round/ (accessed 20 November 2020).]
5 T. Yarwood and F. Castle, *Physical and Mathematical Tables*, London: Macmillan, 1970.
6 D.G. Andrews, J.R. Holton and C.B. Leovy, *Middle Atmosphere Dynamics*, London: Academic Press, 1987.
7 G.M. Dunnavan and J.W. Dierks, 'An Analysis of Super Typhoon Tip', *Monthly Weather Review*, vol. 108, no. 11 (1980), pp. 1915–23.
8 R. Scherhag, 'Die explosionsartigen Stratosphärenerwärmungen des Spät-winters 1951/52', *Ber. Deut. Wetterdieuste*, vol. 6 (1952), pp. 51–63.
9 M. McIntyre and T. Palmer, 'Breaking Planetary Waves in the Stratosphere', *Nature*, vol. 305 (1983), pp. 593–600.

10 M. Baldwin and T. Dunkerton, 'Propagation of the Arctic Oscillation from the Stratosphere to the Troposphere', *Journal of Geophysical Research Atmospheres*, vol. 104, no. 24 (1999), pp. 30937–46.

11 D. Stephenson, H. Wanner, S. Brönnimann and J. Luterbacher, 'The History of Scientific Research on the North Atlantic Oscillation', in *The North Atlantic Oscillation: Climatic Significance and Environmental Impact*, American Geophysical Union, 2003, pp. 37–50.

12 M.P.Baldwin,D.B.Stephenson,D.W.Thompson,T.J.Dunkerton,A.J. Charlton and A. O'Neill, 'Stratospheric Memory and Skill of Extended-Range Weather Forecasts', *Science*, vol. 301, no. 5633 (2003), pp. 636–40.

Chapter 9: Change

1 J. Needham, *Science and Civilisation in China: Volume 3, Mathematics and the Sciences of the Heavens and the Earth*, Taipei: Caves Books, 1986.

2 A. Alexander, *Infinitesimal: How a Dangerous Mathematical Theory Shaped the Modern World*, London: Oneworld Publications, 2014.

3 'James Hutton: Father of Modern Geology 1726–1797', *Nature,* vol. 119 (1927), p. 582.

4 G. Davies, 'Early Discoverers XXVI: Another Forgotten Pioneer of the Glacial Theory, James Hutton (1726–97)', *Journal of Glaciology*, vol. 7, no. 49 (1968), pp. 115–16.

5 E. Evans, 'The Authorship of the Glacial Theory', *The North American Review*, vol. 145 (1887), pp. 94–7.

6 T. Krüger, *Discovering the Ice Ages: International Reception and Consequences for a Historical Understanding of Climate*, Leiden: Brill, 2013.

7 D. Waltham, *Lucky Planet: Why Earth is Exceptional – and What That Means for Life in the Universe*, London: Icon Books, 2014.

8 ibid.

9 A. Chodos, 'March 21, 1768: Birth of Jean-Baptiste Joseph Fourier', American Physical Society, March 2010. [Online: available at https://www.aps.org/publications/apsnews/201003/physicshistory.cfm (accessed 20 November 2020).]

10 S. Blundell and K. Blundell, *Concepts in Thermal Physics*, Oxford: Oxford University Press, 2010.

11 F. Arago, *Biographies of Distinguished Scientific Men*, Boston, 1857.

12 S. Weart, *The Discovery of Global Warming*, Cambridge, MA: Harvard University Press, 2008.

13 E. Foote, 'Circumstances Affecting the Heat of the Sun's Rays', *The American Journal of Science and Arts*, vol. 22 (1856), pp. 382–3.

14 Smithsonian Institution Archives, 'An Act to Establish the "Smithsonian Institution", for the Increase and Diffusion of Knowledge Among Men'. [Online: available at https://siarchives. si.edu/collections/siris_sic_4026 (accessed 20 November 2020).]

15 J. Golinski, 'Enlightenment Science', in *The Oxford Illustrated History of Science*, Oxford, Oxford University Press, 2017, pp. 180–212.

16 J.M. Wallace and P.V. Hobbs, *Atmospheric Science: An Introductory Survey*, Academic Press, 2006.

17 A. Horvitz, S. Stephens, M. Helfert, G. Goodge, K.T. Redmond, K. Pomeroy and E. Kurdy, 'A National Temperature Record at Loma, Montana', in *12th Symposium on Meteorological Observations and Instrumentation*, Long Beach, CA, 2003.

18 N. Ekholm, 'On the Variations of the Climate of the Geological and Historical Past and their Causes', *Quarterly Journal of the Royal Meteorological Society*, vol. 27 (1901).

19 J. Shakun, P. Clark, F. He, S. Marcott, A. Mix, Z. Liu, B. Otto-Bliesner, A. Schmittner and E. Bard, 'Global Warming Preceded by Increasing Carbon Dioxide Concentrations during the Last Deglaciation', *Nature*, vol. 484 (2012), pp. 49–54.

20 J. Neumann, 'Climatic Change as a Topic in the Classical Greek and Roman literature', *Climatic Change*, vol. 7 (1985), pp. 441–54.

21 A. Wulf, *The Invention of Nature: Alexander von Humboldt's New World*, New York: Knopf, 2015.

22 R. Krulwich, 'The Fantastically Strange Origin of Most Coal on Earth', *National Geographic*, 7 January 2016. [Online: available at https://www.nationalgeographic.com/science/article/the-fantastically-strange-origin-of-most-coal-on-earth/ (accessed 9 November 2020).]

23 P. Ward and J. Kirschvink, *A New History of Life*, London: Bloomsbury, 2016.

24 K. Davids and C. Davids, *Religion, Technology, and the Great and Little Divergences: China and Europe Compared, c. 700–1800*, Leiden, Brill, 2012.

25 M. Csele, 'The Newcomen Steam Engine'. [Online: available at http://www.technology.niagarac.on.ca/people/mcsele/interest/the-newcomen-steam-engine/ (accessed 9 November 2020).]

26 E. Roll, *An Early Experiment in Industrial Organisation: Being a History of the Firm of Boulton & Watt, 1775–1805*, Longmans, Green and Co., 1930.

27 P. Ackroyd, *London: The Biography*, London: Chatto & Windus, 2000.

28 S. Weart, *The Discovery of Global Warming*, Cambridge, MA: Harvard University Press, 2008.

29 S. Arrhenius, 'On the Influence of Carbonic Acid in the Air upon the Temperature of the Ground', *The London, Edinburgh, and Dublin Philosophical Magazine and Journal of Science*, vol. 41, no. 5 (1896), pp. 237–76.

30 S. Arrhenius, *Worlds in the Making*, Leipzig: Academic Publishing House, 1908.

31 R. Kunzig and W. Broecker, *Fixing Climate: The Story of Climate Science – and How to Stop Global Warming*, London: Sort Of Books, 2008.

32 D. Harris, 'Charles David Keeling and the Story of Atmospheric CO_2 Measurements', *Analytical Chemistry*, vol. 82, no. 19 (2010), pp. 7865–70.

33 Conservation Foundation, 'Implications of Rising Carbon Dioxide Content of the Atmosphere', New York, 1963.

34 National Academy of Sciences, Committee on Atmospheric Sciences Panel on Weather and Climate Modification, 'Weather and Climate Modification: Problems and Prospects', Washington, DC, 1966.

35 M. Kovenock and A. Swann, 'Leaf Trait Acclimation Amplifies Simulated Climate Warming in Response to Elevated Carbon Dioxide', *Global Biogeochemical Cycles*, vol. 32, no. 10 (2018), pp. 1437–48.

36 M. Mann, *The Hockey Stick and the Climate Wars*, New York: Columbia University Press, 2014.

37 S. Weart, *The Discovery of Global Warming*, Cambridge, MA: Harvard University Press, 2008.

38 N. Oreskes and E. Conway, *Merchants of Doubt: How a Handful of Scientists Obscured the Truth on Issues from Tobacco Smoke to Global Warming*, New York: Bloomsbury Press, 2010.

39 GISTEMP Team, 'GISS Surface Temperature Analysis (GISTEMP) Version 4', NASA Goddard Institute for Space Studies, 2020. [Online: available at https://data.giss.nasa.gov/gistemp/ (accessed 11 November 2020).]

40 D. Wallace-Wells, *The Uninhabitable Earth: A Story of the Future*, London: Penguin, 2019.

41 Royal Meteorological Society, 'The Pliocene: The Last Time Earth had >400 ppm of Atmospheric CO_2', London, 2019.
42 J. Tollefson, 'How Hot Will Earth Get By 2100?', *Nature*, 22 April 2020. [Online: available at · https://www.nature.com/articles/d41586-020-01125-x (accessed 12 November 2020).]
43 REN21, 'Renewables 2020 Global Status Report', Paris, 2020.

Epilogue: Family

1 NASA Exoplanet Science Institute, 'NASA Exoplanet Archive', 2020. [Online: available at https://exoplanetarchive.ipac.caltech.edu/cgi-bin/TblView/nph-tblView?app=ExoTbls&config=PS (accessed 23 November 2020).]
2 D. Armstrong, E. Mooij, J. Barstow, H. Osborn, J. Blake and N.F. Saniee, 'Variability in the Atmosphere of the Hot Giant Planet HAT-P-7 b', *Nature Astronomy*, vol. 1 (2017).
3 J. Yang, N. Cowan and D. Abbot, 'Stabilizing Cloud Feedback Dramatically Expands the Habitable Zone of Tidally Locked Planets', *The Astrophysical Journal Letters*, vol. 771, no. 2 (2013).

ACKNOWLEDGEMENTS

This book would not have been possible without the help of a great many people. Thanks in particular must go to Dr Ryan Kemp and Dr Nicholas Cole for some very helpful initial conversations, and to Dr Hannah Wakeford for her feedback on exoplanetary atmospheres. Thank you also to Professors Mark Baldwin and David Stephenson, my PhD supervisors, for giving me the opportunity to immerse myself in geophysical fluid dynamics, and to Professor Lesley Gray for making my PhD position possible.

The book was written during the 2020 global pandemic, and was in fact delayed because I caught coronavirus. As such, thanks must go to Ian Wong and Huw Armstrong at Hodder & Stoughton and Chris Wellbelove at Aitken Alexander for their continued patience with me. Perhaps as they had nothing better to do during lockdown, I also very gratefully received comments on the manuscript from Charlotte Connelly, Claire Beazley, Shamini Bundell, Ali Jennings, Emily Bates, Simon Lee, Rohin Francis, David Arnold, and Tom Dowling that were extremely helpful.

I would be remiss to not also mention my online community, on YouTube, Twitch, and Discord, for giving me the opportunity to write this book. Your unending support for my work motivates me every single day, and I hope this end product justifies the relative lack of videos in 2020. To the admins and mods in particular, thank you so much for all that you do. All hail Claude.

Works that were particularly useful in the writing of this book include *Jet Stream* by Tim Woollings, *Lucky Planet* by David Waltham, *Don't Even Think About It* by George Marshall, *History of the Meteorological Office* by Malcolm Walker, *Concepts in Thermal Physics* by Stephen Blundell and Katherine Blundell, and *The Uninhabitable Earth* by David Wallace-Wells. Any piece of non-fiction is only as good as the research material available, and I had the pleasure of an extensive literature to draw from. Please refer to the bibliography for a complete list. In particular, if you found the mathematical part of this book interesting, I highly recommend *Atmospheric Science: An Introductory Survey* by John M. Wallace and Peter Hobbs, and *Essentials of Atmospheric and Oceanic Dynamics* by Geoff Vallis for further reading.

Finally, this book would not have been possible without the unending support of my wife Olivia. I apologise that I spent many months holed up in my office even more than usual, and made you endure innumerable titbits of information from my research when I did emerge. I hope the end result was worth it.

GLOSSARY

Air parcel: a hypothetical object used in atmospheric physics, this is a chunk of atmosphere thermally isolated from its environment such that no heat can flow in or out

Arctic Oscillation: the principal way in which atmospheric pressure varies in the northern hemisphere, effectively a see-saw pattern between the Arctic and the Atlantic/Pacific Ocean basins

Barometer: a device used to measure atmospheric pressure

Blackbody radiation: electromagnetic radiation emitted by all objects in the universe, the quantity being proportional to the fourth power of the object's temperature

Climate: the long-term average of atmospheric conditions such as temperature and chemical concentrations

Climate change: large-scale long-term shifts in climate

Coriolis acceleration: the horizontal acceleration experienced by objects moving on a rotating sphere, this deflects objects eastward when moving poleward

Easterly wind: wind that comes from the east, flowing towards the west

Electromagnetic radiation: waves in the electromagnetic field carrying energy. Examples include visible light, infrared radiation, and ultraviolet light

El Niño: an aperiodic current of warm water off the west coast of South America

Eötvös acceleration: the vertical acceleration experienced by objects moving on a rotating sphere, this deflects objects radially outward (i.e. reducing the gravity they experience) when moving east

Exosphere: the outermost layer of the atmosphere, between around 600 km and 10,000 km above the surface

Front: a boundary separating sections of the atmosphere with notably different conditions

Geocorona: the wisp-like top of the Earth's atmosphere, trailing behind the planet having been stripped away by solar radiation

Geostrophic wind: the flow of air resulting from a balance of pressure gradient forces and Coriolis acceleration

Global warming: the increase in global average temperature caused by increased carbon dioxide concentrations in the atmosphere

Greenhouse effect: the trapping of heat in the Earth's atmosphere by natural and anthropogenic compounds that absorb long wavelengths of electromagnetic radiation

Hadley cell: the large-scale overturning circulation in the Earth's tropics

Heat capacity: how much energy (in joules) a kilogram of a given substance requires for it to increase in temperature by one kelvin

Ionisation: the process of a molecule or atom gaining an electric charge by gaining or losing electrons

IPCC: Intergovernmental Panel on Climate Change

Jet stream: a thin band of fast-moving air

Kármán line: the official start of space, 100 km above the Earth's surface

Keeling Curve: the record of atmospheric CO_2 concentration started in 1958 by Charles Keeling

Lapse rate: the rate at which air decreases in temperature with altitude – the tropopause is defined as having a positive lapse rate, while the stratosphere has a negative lapse rate (note the negative built into the definition)

Mesosphere: also known as the ignorosphere, the layer in the atmosphere between 50 km and 80 km above the surface, home to the aurora and not much else

Meteorology: the study of the atmosphere, mostly focusing on short-term phenomena and their prediction

North Atlantic Oscillation: the see-saw of atmospheric pressure between the Azores and Iceland

Ozone: triatomic oxygen, three oxygen atoms stuck together, commonly found in the stratosphere and very good at absorbing ultraviolet light

Paleoclimate: Earth's climate in the deep past

Photolysis: the process of molecules being split apart by high-energy radiation

Polar vortex: the cyclonic circulation of air around the polar winter-time stratosphere

Proto-globalisation: a period roughly between 1600 and 1800 that saw increased international trade and colonisation, and development of large international corporations

Sounding: multiple measurements of the atmosphere at the same location but different heights above the surface

Southern Oscillation: the see-saw pattern of atmospheric pressure between the Pacific and Indian Oceans

Static stability: the property of an atmosphere with a negative lapse rate, such that temperature increases with altitude, preventing large-scale vertical motion

Stratosphere: the second layer of the atmosphere, from around 10 km to 50 km above the surface

Subtropical high: an area of high pressure just north and south of the equator

Sudden stratospheric warming (SSW): the violent, mid-winter destruction of the stratospheric polar vortex due to anomalous wave activity, associated with large local temperature increases

Thermal wind: the vertical wind shear caused by a horizontal temperature gradient, a combination of geostrophic balance in the horizontal and geostrophic balance in the vertical

Thermometer: a device that measures the absolute temperature of an object on a universal scale

Thermoscope: a device that allows for the relative temperatures of two objects to be established, i.e. if one is warmer or colder than the other

Thermosphere: the penultimate layer of the Earth's atmosphere, from around 80 km to 600 km above the surface

Trade wind: persistent wind from east to west just north and south of the equator

Tropopause: the boundary between the troposphere and stratosphere

Troposphere: the bottom-most layer of the atmosphere, from the surface to approximately 10 km in altitude

Wavelength: characteristic of all waves, the distance between one wave crest and another

Weather: short-term variations in atmospheric conditions, such as temperature, humidity or wind

Westerly wind: wind that comes from the west, flowing towards the east

BIBLIOGRAPHY

Abbe, C., 'Memoir of William Ferrel: 1817–1891', Cambridge, 1892.

Ackroyd, P., *London: The Biography*, London: Chatto & Windus, 2000.

Ahman, L., Kanth, R., Parvaze, S. and Mahdi, S., *Experimental Agrometeorology: A Practical Manual*, Springer, 2017.

Alexander, A., *Infinitesimal: How a Dangerous Mathematical Theory Shaped the Modern World*, London: Oneworld Publications, 2014.

Andrews, D.G., Holton, J.R. and Leovy, C.B., *Middle Atmosphere Dynamics*, London: Academic Press, 1987.

AON, 'Global Catastrophe Recap', 2018.

Arago, F., *Biographies of Distinguished Scientific Men*, Boston, 1857.

Armstrong, D., Mooij, E., Barstow, J., Osborn, H., Blake, J. and Saniee, N.F., 'Variability in the Atmosphere of the Hot Giant Planet HAT-P-7 b', *Nature Astronomy*, vol. 1 (2017).

Arrhenius, S., 'On the Influence of Carbonic Acid in the Air upon the Temperature of the Ground', *The London, Edinburgh, and Dublin Philosophical Magazine and Journal of Science*, vol. 41, no. 5 (1896), pp. 237–76.

Arrhenius, S., *Worlds in the Making*, Leipzig: Academic Publishing House, 1908.

Baldwin, M. and Dunkerton, T., 'Propagation of the Arctic Oscillation from the Stratosphere to the Troposphere', *Journal of Geophysical Research Atmospheres*, vol. 104, no. 24 (1999), pp. 30937–46.

Baldwin, M.P., Stephenson, D.B., Thompson, D.W.J., Dunkerton, T.J., Charlton, A.J. and O'Neill, A., 'Stratospheric Memory and Skill of Extended-Range Weather Forecasts', *Science*, vol. 301, no. 5633 (2003), pp. 636–40.

Berner, R. and Kothavala, Z., 'Geocarb III: A Revised Model of Atmospheric CO_2 over Phanerozoic Time', *American Journal of Science*, vol. 301, no. 2 (2001), pp. 182–204.

Best, N., Havens, R. and LaGow, H., 'Pressure and Temperature of the Atmosphere to 120 km', *Physical Review*, vol. 71, no. 12 (1947), pp. 915–16.

Bjerknes, J., 'Atmospheric Teleconnections from the Equatorial Pacific', *Monthly Weather Review*, vol. 97, no. 3 (1969), pp. 163–72.

Blundell, S. and Blundell, K., *Concepts in Thermal Physics*, Oxford: Oxford University Press, 2010.

Bolton, H.C., *Evolution of the Thermometer 1592–1743*, Easton, PA: The Chemical Publishing Co., 1900.

Braun, W. von, 'Recollections of Childhood: Early Experiences in Rocketry as Told by Werner Von Braun 1963'. [Online: available at https://web.archive.org/web/20090212140739/http://history.msfc.nasa.gov/vonbraun/recollect-childhood.html (accessed 12 July 2020).]

Butchart, N., 'The Brewer–Dobson Circulation', *Reviews of Geophysics* (2014), pp. 157–84.

Buys-Ballot, M., 'Note sur le rapport de l'intensité et de la direction du vent avec les écarts simultanés du baromètre', *Académie des sciences (France) Comptes rendus hebdomadaires*, vol. 45 (1857), pp. 765–8.

Cawkwell, G., *Philip of Macedon*, London: Faber & Faber, 1978.

CBS Chicago, 'It's Official, Chicago Is Colder than Parts of the Arctic, Yukon, and Mars', 30 January 2019. [Online: available at https://chicago.cbslocal.com/2019/01/30/chicago-deep-freeze-colder-than-arctic-yukon-mars-siberia-mount-everest/ (accessed 20 November 2020).]

Chodos, A., 'March 21, 1768: Birth of Jean-Baptiste Joseph Fourier', American Physical Society, March 2010. [Online: available at https://www.aps.org/publications/apsnews/201003/physicshistory.cfm (accessed 20 November 2020).]

Choi, C., 'Strange But True: Earth Is Not round', *Scientific American*, 12 April 2007. [Online: available at https://www.scientificamerican.com/article/earth-is-not-round/ (accessed 20 November 2020).]

Conservation Foundation, 'Implications of rising carbon dioxide content of the atmosphere', New York, 1963.

Coriolis, G., 'Sur les équations du mouvement relatif des systèmes de corps', *Journal de l'École Royale Polytechnique*, vol. 15 (1835), pp. 144–54.

Csele, M., 'The Newcomen Steam Engine'. [Online: available at http://www.technology.niagarac.on.ca/people/mcsele/interest/the-newcomen-steam-engine/ (accessed 9 November 2020).]

Dalrymple, W., *The Anarchy: The Relentless Rise of the East India Company*, New York: Bloomsbury Publishing, 2019.

Danforth, C., 'Chaos in an Atmosphere Hanging on a Wall', 2013. [Online: available at http://mpe.dimacs.rutgers.edu/2013/03/17/chaos-in-an-atmosphere-hanging-on-a-wall/ (accessed 7 October 2020).]

Davids, K. and Davids, C., *Religion, Technology, and the Great and Little Divergences: China and Europe Compared, c. 700–1800*, Leiden: Brill, 2012.

Davies, G., 'Early Discoverers XXVI: Another Forgotten Pioneer of the Glacial Theory, James Hutton (1726–97)', *Journal of Glaciology*, vol. 7, no. 49 (1968), pp. 115–16.

Davis, M., *Late Victorian Holocausts: El Niño Famines and the Making of the Third World*, London: Verso, 2000.

Defoe, D., *The Storm*, London, 1704.

Dunnavan, G.M. and Dierks, J.W., 'An Analysis of Super Typhoon Tip', *Monthly Weather Review*, vol. 108, no. 11 (1980), pp. 1915–23.

Edwards, P., *A Vast Machine: Computer Models, Climate Data, and the Politics of Global Warming*, Cambridge, MA: MIT Press, 2010.

Ekholm, N., 'On the Variations of the Climate of the Geological and Historical Past and their Causes', *Quarterly Journal of the Royal Meteorological Society* (1901), vol. 27.

Evans, E., 'The Authorship of the Glacial Theory', *The North American Review*, vol. 145 (1887), pp. 94–7.

Fagan, B., *Floods, Famines, and Emperors*, London: Pimlico, 2000.

Fahrenheit, D., 'Experimenta et observationes de congelatione aquae in vacuo factae', *Philosophical Transactions of the Royal Society*, vol. 33 (1724), pp. 78–89.

Ferrel, W., 'An Essay on the Winds and Currents of the Oceans', *Nashville Journal of Medicine and Surgery*, 1856.

Ferrel, W., 'On the Effect of the Sun and Moon upon the Rotatory Motion of the Earth', *Astronomical Journal*, vol. 3 (1853).

Ferrel, W., 'The Influence of the Earth's Rotation Upon the Relative Motion of Bodies Near its Surface', *Astronomical Journal*, vol. 5, no. 109 (1859), pp. 97–100.

Fleming, J., *Inventing Atmospheric Science: Bjerknes, Rossby, Wexler, and the Foundations of Modern Meteorology*, Cambridge, MA: MIT Press, 2016.

Foote, E., 'Circumstances Affecting the Heat of the Sun's Rays', *The American Journal of Science and Arts*, vol. 22 (1856), pp. 382–3.

Fort, T., *Under the Weather*, London: Century, 2006.

Frankopan, P., *The Silk Roads*, London: Bloomsbury, 2015.

Friedman, R., *Appropriating the Weather: Vilhelm Bjerknes and the Construction of a Modern Meteorology*, Ithaca, NY: Cornell University Press, 2018.

Frisinger, H., 'Aristotle and his "Meteorologica"', *Bulletin of the American Meteorological Society*, vol. 53, no. 7 (1972), pp. 634–8.

Frisinger, H., *The History of Meterology: To 1800*, New York: Science History Publications, 1977.

GISTEMP Team, 'GISS Surface Temperature Analysis (GISTEMP) Version 4', NASA Goddard Institute for Space Studies, 2020. [Online: available at https://data.giss.nasa.gov/gistemp/ (accessed 11 November 2020).]

Glaisher, J., *Travels in the Air*, London: Bentley, 1871.

Global Ozone Research and Monitoring Project, 'Scientific Assessment of Ozone Depletion: 2018', WMO, 2018.

Goldstein, B., 'Ibn Muʿādh's Treatise On Twilight and the Height of the Atmosphere', *Archive for History of Exact Sciences*, vol. 17, no. 2 (1977), pp. 97–118.

Golinski, J., 'Enlightenment Science', in *The Oxford Illustrated History of Science*, Oxford, Oxford University Press, 2017, pp. 180–212.

Götz, F., Meetham, A. and Dobson, G., 'The Vertical Distribution of Ozone in the Atmosphere', *Proceedings of the Royal Society of London. Series A, Containing Papers of a Mathematical and Physical Character*, vol. 145, no. 855 (1934), pp. 416–46.

Gregory, A., *Eureka! The Birth of Science*, London: Icon Books, 2001.

Hadley, G., 'Concerning the Cause of the General Trade Winds', *Philosophical Transaction of the Royal Society of London*, vol. 39, no. 437 (1735), pp. 58–62.

Harris, D., 'Charles David Keeling and the Story of Atmospheric CO_2 Measurements', *Analytical Chemistry*, vol. 82, no. 19 (2010), pp. 7865–70.

Hellmann, G., 'The Dawn of Meteorology', *Quarterly Journal of the Royal Meteorology Society*, vol. 34, no. 148 (1908).

Hickel, J., 'How Britain Stole \$45 Trillion from India', 19 December 2018. [Online: available at https://www.aljazeera.com/opinions/2018/12/19/how-britain-stole-45-trillion-from-india/ (accessed 18 October 2020).]

Holland, H., 'The Oxygenation of the Atmosphere and Oceans', *Philosophical Transactions of the Royal Society of London B: Biological Sciences*, vol. 361, no. 1470 (2006), pp. 903–15.

Holmes, R., *Falling Upwards: How We Took to the Air*, London: William Collins, 2013.

Hopkins, A., *Globalization in World History*, New York: Norton, 2002.

Horvitz, A., Stephens, S., Helfert, M., Goodge, G., Kelly, T.R., Pomeroy, K. and Kurdy, E., 'A National Temperature Record at Loma, Montana', in *12th Symposium on Meteorological Observations and Instrumentation*, Long Beach, CA, 2003.

Howes, A., 'Age of Invention: The Spanish Engine', 24 July 2020. [Online: available at https://antonhowes.substack.com/p/age-of-invention-the-spanish-engine (accessed 9 November 2020).]

Huygens, C., *Oeuvres completes de Christiaan Huygens publiees par la Societe Hollandaise des Sciences*, The Hague, 1893.

'James Hutton: Father of Modern Geology 1726–1797', *Nature*, vol. 119 (1927), p. 582.

JAXA, 'Research on Balloons to Float over 50 km Altitude', Institute of Space and Astronautical Science, 2008. [Online: available at http://www.isas.jaxa.jp/e/special/2003/yamagami/03.shtml (accessed 18 November 2020).]

Kottek, M., Grieser, J., Beck, C., Rudolf, B. and Rubel, F., 'World Map of the Köppen-Geiger Climate Classification Updated', *Meteorologische Zeitschrift*, vol. 15, no. 3 (2006), pp. 259–63.

Kovenock, M. and Swann, A., 'Leaf Trait Acclimation Amplifies Simulated Climate Warming in Response to Elevated Carbon Dioxide', *Global Biogeochemical Cycles*, vol. 32, no. 10 (2018), pp. 1437–48.

Krüger, T., *Discovering the Ice Ages. International Reception and Consequences for a Historical Understanding of Climate*, Leiden: Brill, 2013.

Krulwich, R., 'The Fantastically Strange Origin of Most Coal on Earth', *National Geographic*, 7 January 2016. [Online: available at https://www.nationalgeographic.com/science/article/the-fantastically-strange-origin-of-most-coal-on-earth/ (accessed 9 November 2020).]

Kumar, K., Rajagopalan, B., Hoerling, M., Bates, G. and Cane, M., 'Unraveling the Mystery of Indian Monsoon Failure During El Niño', *Science*, vol. 314, no. 5796 (2006), pp. 115–19.

Kunzig, R. and Broecker, W., *Fixing Climate: The Story of Climate Science – and How to Stop Global Warming*, London: Sort Of Books, 2008.

Lehman, M. and Goddard, Robert H., *Pioneer of Space Research*, New York: Da Capo Press, 1988.

Lewis, J., 'Ooishi's Observation Viewed in the Context of Jet Stream Discovery', *Bulletin of the American Meteorological Society*, vol. 84, no. 3 (2003), pp. 357–70.

Lindemann, F.A. and Dobson, G.M.B., 'A Theory of Meteors, and the Density and Temperature of the Outer Atmosphere to which it Leads', *Proceedings of the Royal Society of London. Series A, Containing Papers of a Mathematical and Physical Character*, vol. 102, no. 717 (1923), pp. 411–37.

Lorenz, E., 'Deterministic Nonperiodic Flow', *Journal of the Atmospheric Sciences*, vol. 20, no. 2 (1963), pp. 130–41.

Lorenz, E., *The Essence of Chaos*, London: UCL Press, 1993.

Mann, M., *The Hockey Stick and the Climate Wars*, New York: Columbia University Press, 2014.

Matsumi, Y. and Kawasaki, M., 'Photolysis of Atmospheric Ozone in the Ultraviolet Region', *Chemical Review*, vol. 103, no. 12 (2003), pp. 4767–82.

McIntyre, M. and Palmer, T., 'Breaking Planetary Waves in the Stratosphere', *Nature*, vol. 305 (1983), pp. 593–600.

Moore, P., *The Weather Experiment: The Pioneers who Sought to See the Future*, London: Chatto & Windus, 2015.

Napier Shaw, W., *Manual of Meteorology*, Cambridge: Cambridge University Press, 1926.

NASA Exoplanet Science Institute, 'NASA Exoplanet Archive', 2020. [Online: available at https://exoplanetarchive.ipac.caltech.edu/cgi-bin/TblView/nph-tblView?app=ExoTbls&config=PS (accessed 23 November 2020).]

National Academy of Sciences, Committee on Atmospheric Sciences Panel on Weather and Climate Modification, 'Weather and Climate Modification: Problems and Prospects', National Academy of Sciences, Washington, DC, 1966.

National Hurricane Center, 'Hurricane Elena Preliminary Report', National Oceanic and Atmospheric Administration, Miami, 1985.

Needham, J., *Science and Civilisation in China: Volume 3, Mathematics and the Sciences of the Heavens and the Earth*, Taipei: Caves Books, 1986.

Neuenschwander, D., *Emmy Noether's Wonderful Theorem*, Baltimore, MD: Johns Hopkins University Press, 2017.

Neufeld, M., *Von Braun: Dreamer of Space, Engineer of War*, New York: A.A. Knopf, 2007.

Neufeld, M.J., *The Rocket and the Reich: Peenemünde and the Coming of the Ballistic Missile Era*, New York: The Free Press, 1995.

Neumann, J., 'Climatic Change as a Topic in the Classical Greek and Roman Literature', *Climatic Change*, vol. 7 (1985), pp. 441–54.

NOAA, 'Hurricanes: Frequently Asked Questions', 1 June 2021. [Online: available at https://www.aoml.noaa.gov/hrd-faq/ (accessed 20 July 2021).]

Nutman, A., Bennett, V., Friend, C., van Kranendonk, M. and Chivas, A., 'Rapid Emergence of Life Shown by Discovery of 3,700-million-year-old Microbial Structures', *Nature*, vol. 537, no. 7621 (2016), pp. 535–8.

O'Connor, J. and Robertson, E., 'Gaspard Gustave de Coriolis', MacTutor, July 2000. [Online: available at https://mathshistory. st-andrews.ac.uk/Biographies/Coriolis/ (accessed 5 October 2020).]

Oestreicher, C., 'A History of Chaos Theory', *Dialogues in Clinical Neuroscience*, vol. 9, no. 3 (2007), pp. 279–89.

Oreskes, N. and Conway, E., *Merchants of Doubt: How a Handful of Scientists Obscured the Truth on Issues from Tobacco Smoke to Global Warming*, New York: Bloomsbury Press, 2010.

Peterson, T., Connolley, W. and Fleck, J., 'The Myth of the 1970s Global Cooling Scientific Consensus', *Bulletin of the American Meteorological Society*, vol. 89, no. 9 (2008), pp. 1325–38.

Ramsey, S., *Tools of War: History of Weapons in Early Modern Times*, Delhi: Vij Books, 2016.

Rasmussen, S., 'Advances in 13th Century Glass Manufacturing and their Effect on Chemical Progress', *Bulletin for the History of Chemistry*, vol. 33, no. 1 (2008), pp. 28–34.

Redfield, W., 'Remarks on the Prevailing Storms of the Atlantic Coast, of the North American States', offprint from *The American Journal of Science and Arts*, vol. 20 (1831), pp. 17–51.

REN21, 'Renewables 2020 Global Status Report', Paris, 2020.

Richardson, L., *Weather Prediction by Numerical Process*, Cambridge: Cambridge University Press, 1922.

Roll, E., *An Early Experiment in Industrial Organisation: Being a History of the Firm of Boulton & Watt, 1775–1805*, Longmans, Green and Co., 1930.

Ropelewski, C. and Halpert, M., 'Global and Regional Scale Precipitation

Patterns Associated with the El Niño/Southern Oscillation', *Monthly Weather Review*, vol. 115 (1987), pp. 1606–26.

Royal Meteorological Society, 'The Pliocene: The Last Time Earth had >400 ppm of Atmospheric CO_2', London, 2019.

Royer, D., Berner, R., Montañez, I., Tabor, N. and Beerling, D., 'CO_2 as a Primary Driver of Phanerozoic Climate', *GSA Today*, vol. 14, no. 3 (2004).

Scherhag, R., 'Die explosionsartigen Stratosphärenerwärmungen des Spät-winters 1951/52', *Ber. Deut. Wetterdieuste*, vol. 6 (1952), pp. 51–63.

Schirrmeister, B., de Vos, J., Antonelli, A. and Bagheri, H., 'Evolution of Multicellularity Coincided with Increased Diversification of Cyanobacteria and the Great Oxidation Event', *Proceedings of the National Academy of Sciences of the United States of America*, vol. 110, no. 5 (2013), pp. 1791–6.

Seilkopf, H., 'Maritime Meteorologie: Vol 2', in *Handbuch der Fliegerwetterkunde*, Radetzke, 1939, p. 359.

Shakun, J., Clark, P., He, F., Marcott, S., Mix, A., Liu, Z., Otto-Bliesner, B., Schmittner, A. and Bard, E., 'Global Warming Preceded by Increasing Carbon Dioxide Concentrations during the Last Deglaciation', *Nature*, vol. 484 (2012), pp. 49–54.

Sheppard, P., 'Obituary of Sir Gilbert Walker, CSI, FRS', *Quarterly Journal of the Royal Meteorological Society*, vol. 83, no. 364 (1959).

Singh, K., *I Shall Not Hear the Nightingale*, New York: Grove Press, 1959.

Smithsonian Institution Archives, 'An Act to Establish the "Smithsonian Institution", for the Increase and Diffusion of Knowledge Among Men'. [Online: available at https://siarchives.si.edu/collections/siris_sic_4026 (accessed 20 November 2020).]

Sobel, D., *Longitude: The True Story of a Lone Genius Who Solved the Greatest Scientific Problem of His Time*, London: Harper Perennial, 2011.

Staff Members of the Department of Meteorology, 'On the General Circulation of the Atmosphere in Middle Latitudes', *Bulletin of the American Meteorological Society*, vol. 28 (1947), pp. 255–80.

Stephenson, D., Wanner, H., Brönnimann, S. and Luterbacher, J., 'The History of Scientific Research on the North Atlantic Oscillation', in *The North Atlantic Oscillation: Climatic Significance and Environmental Impact*, American Geophysical Union, 2003, pp. 37–50.

Strogatz, S., *Nonlinear Dynamics and Chaos: With Applications to Physics,*

Biology, Chemistry, and Engineering, Boulder, CO: Westview Press, 2015.

Teague, K. and Gallicchio, N., *The Evolution of Meteorology: A Look in the Past, Present, and Future of Weather Forecasting*, Oxford: Wiley, 2017.

Tollefson, J., 'How Hot Will Earth Get By 2100?', *Nature*, 22 April 2020. [Online: available at https://www.nature.com/articles/d41586-020-01125-x (accessed 12 November 2020).]

Tyrell, K., 'Oldest Fossils Ever Found Show Life on Earth Began Before 3.5 Billion Years Ago', University of Wisconsin-Madison, 18 December 2017. [Online: available at https://news.wisc.edu/oldest-fossils-found-show-life-began-before-3-5-billion-years-ago/ (accessed 2 November 2020).]

UK Met Office, 'Global Accuracy at a Local Level'. [Online: available at https://www.metoffice.gov.uk/about-us/what/accuracy-and-trust/how-accurate-are-our-public-forecasts (accessed 20 July 2021).]

UK Met Office, 'What is Saharan Dust?' [Online: available at https://www.metoffice.gov.uk/weather/learn-about/weather/types-of-weather/wind/saharan-dust (accessed 20 July 2021).]

UK Parliament, 'The Witchcraft Act', 1735.

University of Warwick, '5400mph Winds Discovered Hurtling Around Planet Outside Solar System', 13 November 2015. [Online: available at https://warwick.ac.uk/newsandevents/pressreleases/5400mph_winds_discovered/ (accessed 4 August 2021).]

Vallis, G., *Atmospheric and Oceanic Fluid Dynamics*, Cambridge: Cambridge University Press, 2006.

Vigh, J., 'Formation of the Hurricane Eye', in *27th Conference on Hurricanes and Tropical Meteorology*, Monterey, American Meteorological Society, 2006.

Wainwright, G., *The Sky Religion in Egypt*, Cambridge: Cambridge University Press, 1938.

Walker, M., *History of the Meteorological Office*, Cambridge: Cambridge University Press, 2012.

Wallace, J.M. and Hobbs, P.V., *Atmospheric Science: An Introductory Survey*, Boston: Academic Press, 2006.

Wallace-Wells, D., *The Uninhabitable Earth: A Story of the Future*, London: Penguin, 2019.

Waltham, D., *Lucky Planet: Why Earth Is Exceptional – and What That Means for Life in the Universe*, London: Icon Books, 2014.

Wang, C., Deser, C., Yu, J.-Y., DiNezio, P. and Clement, A., 'El Niño and Southern Oscillation (ENSO): A Review', in *Coral Reefs of the Eastern Pacific*, Dordrecht: Springer Science, 2016, pp. 85–106.

Ward, P. and Kirschvink, J., *A New History of Life*, London: Bloomsbury, 2016.

Weart, S., *The Discovery of Global Warming*, Cambridge, MA: Harvard University Press, 2008.

West, J., 'Torricelli and the Ocean of Air: The First Measurement of Barometric Pressure', *Physiology*, vol. 28, no. 2 (2013), pp. 66–73.

Whipple, F., 'The Propagation of Sound to Great Distances', *Quarterly Journal of the Royal Meteorological Society*, vol. 61, no. 261 (1935).

Wilkinson, R., *The Complete Gods and Goddesses of Ancient Egypt*, London: Thames & Hudson, 2003.

'William C. Redfield 1789–1857', *Weatherwise*, vol. 22, no. 6 (1969), pp. 225–62.

Woollings, T., *Jet Stream: A Journey Through Our Changing Climate*, Oxford: Oxford University Press, 2019.

Wulf, A., *The Invention of Nature: Alexander von Humboldt's New World*, New York: Knopf, 2015.

Yan, Y., Bender, M., Brook, E., Clifford, H., Kemeny, P., Kurbatov, A., Mackay, S., Mayewski, P., Ng, J., Sveringhaus, J. and Higgins, J., 'Two-million-year-old Snapshots of Atmospheric Gases from Antarctic Ice', *Nature*, vol. 574 (2019), pp. 663–6.

Yang, J., Cowan, N. and Abbot, D., 'Stabilizing Cloud Feedback Dramatically Expands the Habitable Zone of Tidally Locked Planets', *The Astrophysical Journal Letters*, vol. 771, no. 2 (2013).

Yarwood, T. and Castle, F., *Physical and Mathematical Tables*, London: Macmillan, 1970.

Zahnle, K., Schaefer, L. and Fegley, B., 'Earth's Earliest Atmospheres', *Cold Spring Harbor Perspectives in Biology*, vol. 2, no. 10 (2010).

Zerkle, A. and Mikhail, S., 'The Geobiological Nitrogen Cycle: From Microbes to the Mantle', *Geobiology* (2017), vol. 15, no. 3, pp. 343–52.

Zinszer, H., 'Meteorological Mileposts', *Scientific Monthly*, vol. 58, no. 4 (1944).

INDEX

Abbe, Cleveland 116
ablation zone 150
Adams, Douglas 69
additional warming 162
Adhémar, Joseph 152
adiabatic cooling 69
advection 119
aeronautics 39
aeronauts 1
Agassiz, Louis 150–1
agriculture 6
air 15–16, 17–18
air masses 121, 122
air parcels 52, 68–70, 88, 89, 211
    changing lanes 85
air pressure
    atmosphere 74
    balloons and 31, 66, 68
    changes in 60, 124
    lowering 80, 106, 136
    measured 116–17
    pressure fields 58
    storms 49
    stratosphere 39, 58–9
air quality 170

air temperature
    changing 60, 63
    constant 36
    decreasing 15, 19, 66, 68, 69
    increasing 36–7, 65, 67–8, 70,
        75, 133, 138
    in the stratosphere 36
al-Biruni, Abu Rayhan 149
albedo 153, 171, 179
Ancient Greek natural philosophy
    10
angular momentum 86, 87, 88
Anticyclone Hartmut 131
anticyclone systems 134
apsidal precession 152, 153
Arctic Oscillation 211
Aristotle 8–10, 11, 16, 46, 73, 80,
    149
Arrhenius, Svante 170–2, 179
Asiatic 76
Assmann, Richard 32–3, 96
astrometeorology 7, 46
astronautics 39
astronomy 9
Atlantic Ocean 80, 109

atmosphere xiii–xvii, 6
absorption of high-frequency solar
    radiation 37
  blackbody radiation 63, 66
  colder at altitude 18–19
  equation of state 58–60, 65, 69
  gods and 6–7
  height of 28, 30, 39–40
  as an insulator 158–9
  interest and speculation 6
  layers 31–2, 36, 37
  measurements of 10–11, 33
  Meteorologica (Aristotle) 8–10
  origins of 22
  soundings of 30–1, 32
  speed of 84
  thinning at altitude 18–19
  variations in 27
  vertical structure of 38
  water in 159
atmospheric engine 168
atmospheric explosion 138, 139
atmospheric physics 59, 121,
    122
atmospheric science xv, 8, 81, 82,
    193–5
  Galileo and 12–13
  origins of 94
atmospheric trough 76
atmospheric waves 140–1, 142
Ayanz y Beaumont, Jerónimo de
    167–8

Babylonians 7
bacteria 166
Baldwin, Mark 143, 144
balloons 1, 2–6, 30–1, 33, 66
  air pressure and 31, 66, 68
  hydrogen-filled 97–8

bamboo 147
barometers 10–11, 16–18, 211
Beast from the East 131, 145
beetles 154
Bergen School of Meteorology
    121, 122–3
Berlin warming 138
Berner, Robert 25
Bernoulli, Daniel 49
Bible 149
Bjerknes, Carl Anton 119
Bjerknes, Jacob 104–5, 119
Bjerknes, Vilhelm 119–20, 121
blackbody radiation 61–2, 156,
    ·211
  of the atmosphere 63, 66
  of the Earth 63, 66
blocking highs 99
boulders 149, 150
Boyle, Robert 14
Brewer-Dobson circulation 70
British East India Company 78,
    81, 101, 194
British Empire 101–2
Bruno, Giordano 189
Buys Ballot, Christophorus
    Henricus Diedericus 53–4
Buys Ballot's law 53–4

Cabot, John 80
Cambrian Period 25
carbon 23, 166–7
carbon dioxide 157, 161–3, 164,
    170
  concentrations of 174–9,
    181–2, 184, 185
carbonic acid 163
Carboniferous Period 166
Catholic Church 16, 149

Celsius, Anders 14–15
Celsius scale 61
centrifugal acceleration 90, 91, 135
chaos theory 128–30
Charles II, King 78
Charney, Jule 124
Chicago 131
Chicago Department 96
chlorofluorocarbons (CFCs) 67
Christy, Miller 132
Circular Theory 49, 52
climate 24–5, 211
    definition 93, 148
    duration 148
    temperature swings 154
    weather and 93, 147–8
climate change xiv–xv, xvii, 148, 149–87, 182, 211
    incremental change 182
    local impact 182
    predictions of 185–6
    uneven distribution of impact 182–3
climatology 78
cloud types 122
coal 166, 169–70
Coffin, James Henry 51
cold fronts 121–2
collisions 87
colonisers
    British 101–2
    Spanish 102–3, 106
Columbus, Christopher 80
computer models 123–5, 128
    complexity of 125
    stratospheric information 145–6
    tropospheric information 145–6

'Concerning the Cause of the General Trade Winds' (Hadley) 84
convection 65, 69–70, 75
Copernicus, Nicolaus 16
Coriolis acceleration 88–9, 109, 117, 211
Coriolis deflection 52
Coriolis, Gaspard-Gustave 88
correlation 103
Coxwell, Henry 1, 2, 3–6, 19, 30, 66, 70, 71–2
critical surface 142
Croll, James 151–3, 154, 164, 173, 193
cumulonimbus clouds 71
Curie, Marie 158
currents 44, 102, 104, 105
    see also El Niño
Cyclone Olivia 44
cyclones 48–9

Darwin, Charles 112, 113, 116
data 81, 82–3
De Historia Piscium (Willughby) 77
Defoe, Daniel 46, 47
deforestation 165
deterministic equations 126, 127, 128
'Deterministic Nonperiodic Flow' (Lorenz) 127
The Discovery of Global Warming (Weart) 181
distance 93–108
domains 56
dripping paint plot 143
droughts 102, 104, 106, 107
du Châtelet, Émilie 158

Dunkerton, Tim 143, 144
dynamic fields 57, 59

Earth
    axis 51, 75, 84–5, 90, 134
    blackbody radiation 63, 66
    calculation of warmth 155–6
    as centre of the universe 9
    circumference of 135
    energy transmission 62
    first atmosphere of 22
    as an oblate spheroid 135
    orbiting the sun 16, 152–3
    polar circumference of 135
    reshaping of life on 24
    rotation of 50, 51–2, 84, 86,
        90, 135
    Sun and 75
    variation in temperature
        159–60
    wavelengths 62–3
earthlight 63
east-west motion 89–90
easterly wind 84, 136, 211
eccentricity of orbit 152
eddy-driven jet stream 95
Ekholm, Nils 160, 161
El Niño 102–3, 107, 108, 211
    early writings about 102
    Walker circulation 106
    worldwide impact of 103
El Niño Southern Oscillation
        (ENSO) 104–5, 107, 146
electromagnetic radiation 60–1,
        211
electromagnetic spectrum 62
electrons 24
elements 23
elliptical orbit 152

energy
    kinetic 13, 88
    splitting of 63–4
    thermal 67
    transfer of 60–1, 62, 63, 74–5
ensemble forecasts 129
Eötvös acceleration 91, 211
Eötvös, Baron Roland von 91
equation of state 57–60, 65, 69
equations 57–72, 104
    'On the Equations of Relative
        Movement of Systems of
        Bodies' (Coriolis) 89
equator 33, 74–5, 84–5, 135–6
equilibrium 55–6
erratics 149, 150
Esperanto 97
The Essence of Chaos (Lorenz)
        126
Etesians 73, 74, 76
Euler, Leonhard 189, 190
evaporation 26, 106
exoplanets 189, 190
exosphere 38, 40, 212
'The Explosive Stratospheric
        Warming of Late Winter
        1951/52' (Scherhag) 138

Fagan, Brian 107
Fahrenheit, Daniel Gabriel 14
famines 101, 104
female scientists 158
Ferdinando II de' Medici, Grand
        Duke of Tuscany 14
Ferranti Mk I 124
Ferrel, William 49–52, 53, 60,
        110, 116, 173, 193
fields 55–72, 118
    equations 57–72

final warming 139
FitzRoy, Robert 112–16, 195
flooding 8
flow 45–6, 51
fluid 44–5
fluid flow 49, 51, 52–3
Foote, Eunice Newton 157–8, 159, 193
forecasts *see* weather forecasts
fossil fuels 186
fossils 23
Fourier, Jean-Baptiste Joseph 155–6, 158, 159, 161
fourth power relationship 61
Franklin, Benjamin 47
freezing weather 131
Fritsch, Ferdinand Helfreich 189, 190
fronts 121–3, 212
Fu-Go balloons 97–8

Galileo Galilei 11–14, 15–16, 46
geochemistry 173–4
geocorona 40, 212
geological carbon cycle 163
geology 149, 150
geophysical fluid dynamics (GFD) 50–1
geostrophic approximation 117
geostrophic wind 117, 118, 212
glaciers 149–50
Glaisher, James 1–6, 19, 30, 66, 70, 71–2, 111, 194–5
glass 12
global carbon cycle 164
global cooling 180
global warming 178–81, 182–3, 185, 212
global wind patterns 79

Goddard, Robert Hutchings 34, 35
gravitational acceleration 135
gravity 90, 91
Great Flood 149
Great Lakes freeze 145
greenhouse effect 159, 160–1, 212
grid size 123

Hadley cell 85, 212
Hadley, George 83–6, 88
Halley, Edmond 76–80, 83
Hamilton, Margaret 125–6
Hansen, James 180
HAT-P-7b (planet) 191–2
heat capacity 212
heat, propagation of 155
heavens, study of 6–7
Henry, Professor Joseph 111, 157–8
Herschel, Caroline 158
Hertz, Henrich 120
'An Historical Account of the Trade Winds.' (Halley) 78–9, 80
HMS *Beagle* 112, 113
Högbom, Arvid 170
Holmes, Richard 2
horizontal pressure gradients 116
hot air balloons 1, 2–6
hot Jupiters 190
Howes, Anton 168
Humboldt, Alexander von 165
Humboldtian science 111, 165, 194
Hurricane Elena 109–10, 128, 129–30, 145
hurricanes 48, 109

Hutton, James 150
Huygens, Christiaan 14
hydrogen-filled balloons 96, 97–8
hypotheses 81

Ibn Al-Haytham (Alhazen) 28, 29
Ibn Muʿādh al-Jayyānī 28–9, 30
Ibn-Sina 149
Ice Age 150–1, 152–3
ice ages 153, 162, 171
ice albedo feedback theory 153,
     171, 179
ice cores 153–4, 161
ice formation 152–3
ice-free summers 153
ice sheets 21, 150, 153, 155
ideal gas law 59, 134
'Implications of Rising Carbon
     Dioxide Content of the
     Atmosphere' (Keeling et al.)
     178
India 100–1
Indian Ocean 100, 103, 105, 106
indivisibles, theory of 16
industrial power 165–6
Industrial Revolution 169
'The Influence of the Earth's
     Rotation Upon the Relative
     Motion of Bodies Near its
     Surface' (Ferrel) 51–2
infrared radiation 63, 157, 158
initial conditions 126, 128, 129
International Geophysical Year
     175–6
interstellar radiation 156
ionisation 37, 212
IPCC (Intergovernmental Panel
     on Climate Change) 180–1,
     183, 212

Iranian low 76
isobars 118
isothermal layer 31–2, 33
isotopes 23, 25

jet stream 94–8, 137, 212
   connection with vortex 145
   discovery of 95–7
   FU-GO balloons 97–8
   global factors 100
   in hemispheres 95
   SSWs and 143
   wayward 144
   weather systems 98–9
Jet Stream: A Journey Through Our
     Changing Climate (Woollings)
     97–8

Kármán line 39, 212
Keeling, Charles David 173–8
Keeling Curve 176–7, 212
kelvin 61
kerosene-soaked paper balloons
     30–1
kinetic energy 13, 88
Kolmogorov, Andrey
     Nikolaevich 128

La Niña 102, 107, 108
Lagrange, Joseph-Louis 49
Laplace, Pierre-Simon 50
lapse rate 19, 31, 212
Lepidodendrales 166
light, opaqueness and transparency
     of 156–7
lignin 166
linear momentum 86, 87
linear velocity 88
Linnaeus, Carl 15

liquid-fuelled rockets 34–5
Llevantades 73
Lorenz, Edward Norton 125–7
Lovelace, Ada 158
Lynch, Peter 124

Mammoth 2, 3–5, 19
Mars 191
mathematics 12
Mauna Loa observatory 175, 176
measured air pressures 116–17
Meitner, Lise 158
Merchants of Doubt (Oreskes and
    Conway) 181
Mersenne, Marin 17
mesosphere 37, 38, 212
Meteorologica (Aristotle) 8–10, 11
meteorological observations
    76–80, 110–11
meteorological offices 111–12,
    114
meteorological unpredictability 94
meteorology 8, 9, 46, 84, 212
    modern 46, 54, 112, 121
    observations 110–11
    theory of 116
Meteorology, Department of
    (University of Chicago) 95
meteors 36, 37
mid-latitude jet stream 95, 96, 99
midnight sun 134
Milanković cycles 154, 162
Milanković, Milutin 153, 154
Mistral 73
modern mechanics 88
modern meteorology 46, 54, 112,
    121
moisture 44, 70–1, 100, 105, 107
molecules 39, 40

momentum 86–7, 141
    conservation of 87
monsoon 100–2, 103
    El Niño 106
    failure 106
    importance of 100, 101
    Southern Oscillation Index
        104
    understanding of 101–2
    Walker circulation 105–6
movingness 86
multicellular life 24
murmuration 45, 48, 51

NACA (National Advisory
    Committee for Aeronautics)
    36
Nagasaki 98
Napoleon 155
National Academy of Sciences
    178
natural philosophers 46, 81
natural philosophy 7, 12, 46
Nazi rocketry programme 35
Neptune 44
Neumann, John von 124
neutrons 23, 24
New World 80–1
Newcomen engine 168–9
Newton, Sir Isaac 46, 50, 51–2,
    77, 126, 128
nitrogen 22
Noether, Emmy 87, 158
North Atlantic Oscillation (NAO)
    103, 144–5, 212
index 144–5
north-south motion 88–9
northern hemisphere 43, 49, 54,
    75

continents and oceans 140, 141–2
jet stream 95, 143
in the summer 75, 100
wave activity 143
in the winter 136, 137, 139
zonally asymmetric 142
numerical weather prediction 124

observation networks 111, 121
observations, meteorological 76–80, 110–11
ocean waves 141
oceans 102, 105–6, 107
Atlantic 93
carbon dioxide 162
Indian 100, 103, 105, 106
Pacific 96, 102, 103, 104, 105, 106, 107, 108
Ooishi, Wasaburo 96–8, 193
opaqueness 156–7
orbital eccentricity 152
overturning circulations 65, 70, 125
oxygen 23–4, 25
oxygen-16 25, 26–7
oxygen-18 25–7
oxygen concentrations 175
Oxygen Holocaust 24
oxygen isotopes 25, 26–7
ozone 37, 63, 67, 156, 157, 212
depletion and recovery 67
heating 67
importance of 67

Pacific Ocean 96, 102, 103, 104, 105, 106, 107, 108

palaeoentomology 154
paleoclimate 147–8, 154, 212
Pascal, Blaise 17–18
Perier, Florin 18
Permian Period 25, 161
Peter the Great 77
Philip II of Macedon 73
Philosophiæ Naturalis Principia Mathematica (Newton) 77
photolysis 212
photosynthesis 23, 166, 174, 175
Physical Review 36
physics 50, 61, 116
planets 60
Pliocene 184, 185
plutonium 98
Poincaré, Henri 128
polar jet stream 95, 96, 99
polar night 134
polar regions 26, 119
hemisphere-scale circulation 134
summer and winter 134
sunlight 134
polar vortex 137, 213
see also stratospheric polar vortex
poles (Earth's) 33, 74–5
Poynting, John Henry 160
prediction 110, 112, 114, 116, 120
see also weather forecasts
pressure fields 58
pressure gradient force 117, 136
primitive equations 120
propagation of heat 155
'The Propagation of Sound to Great Distances' (Whipple) 132

proto-globalisation 81, 82, 213
protons 23, 24

qualitative weather prediction 123
quantitative weather prediction
        123

rain 163
Redfield, William 48–9, 52
Renaissance Italy 11–12
renewable technologies 186–7
'Republic of Letters' 17
resolution, of computers 123, 128
Richardson, Lewis Fry 123–4
rockets 34–6
Rossby, Carl-Gustaf 96, 121
Royal Charter 114, 115
Royal Meteorological Society 132
Royal Society of London 77, 82

Saint Helena 78, 81
Scherhag, Richard 138, 139
Schimper, Karl Friedrich 150
science 7
        data and 81, 82–3, 183
        debt to 82–3
        exclusion of women 158
        self-perception of 82
        societal factors 173
sea levels, rising 184
seasons 133, 134
see-saw of air mass 103–4, 107
Seilkopf, Heinrich 96
Shaw, Sir William Napier 11
Shen Kuo 147, 149, 193
Singh, Khushwant 100
single-celled organisms 22, 23
Sirocco 73
sloshing of air mass 103–4, 107

Smithsonian Institution 157
snow 21–2
solar radiation 152–3, 154, 156
Somerville, Mary 158
sound waves 132
soundings 30–1, 32, 213
southern hemisphere 49
        jet stream 95
        oceans 140
        in the winter 136, 137
        zonally symmetric 142
Southern Oscillation 103–4, 107,
        144, 213
Southern Oscillation Index 104
Spain 102–3
static stability 64, 70, 71, 133,
        213
statistics 103, 104, 144
steam engines 167–70, 195
stellar nebula 22
The Storm (Defoe) 46
storms 48–9, 51, 110
        warnings 114
strategic points of world weather
        103, 144
stratosphere 33, 36, 38, 66–7, 70,
        213
        computer modelling 145–6
        as a ghostly realm 71
        interaction with troposphere
        145
        statically stable 70, 71, 133
stratospheric polar vortex 137
        connection with jet stream
        145
        deceleration of 141, 142–3
        northern 137, 140
        size of 137
        southern 137, 140

SSW and 139
in summer 139
in winter 139
stratospheric surf zone 141
stratospheric winds 134
subtropical high 213
subtropical jet stream 95
sudden stratospheric warming
    (SSW) 138–9, 139–40, 143,
    213
warning signs 145, 146
Sun
    energy transfer 60–1, 63, 74–5
    power output 154
    setting of 28–9
    as a stellar forge 60
sunlight 63
surface gravity 135
synoptic charts 113–14, 115,
    118

Teisserenc de Bort, Léon Philippe
    30–2, 33
teleconnections 146
telegraph 111
temperature 13–15, 74
    global 25
    variations in 159–60
temperature fields 58
temperature gradients 100, 106,
    136
temperature inversion 32
temperature scale 14–15
Thales of Miletus 7–8
The Great Storm (1703) 46–7
Theophrastus 149, 164–5
*Theoria Motuum Planetarum et
    Cometarum* (Euler) 189
thermal energy 67

thermal expansion 13
thermal radiation 156, 157, 159,
    160
thermal wind 136–7, 213
thermometers 10–11, 14–15, 129,
    213
thermoscopes 12–14, 15
thermosphere 37, 38, 39, 213
tides 50
Titan 191
Torricelli, Evangelista 16–17,
    149
trade winds 80–91, 213
    discovery of 80
    Hadley's theory of 84–6, 88
    Halley's theory of 80, 82, 83
    importance of 81
transparency 157
tropical cyclones 109
tropopause 33, 213
troposphere 33, 38, 66, 133,
    135–6, 213
    circulation 134–5
    computer modelling 145–6
    conditional stability 69–70
    interaction with stratosphere
    145
tropospheric polar vortex 137
turbopause 40
turningness 86–7
twilight 28–9
Tyndall, John 156, 158
Typhoon Tip 137

UK Met Office 123, 129
ultraviolet light 63, 67
UNIVAC I 124
universe, Aristotle's concept of
    9–10, 16

University of Strathclyde (Andersonian University) 151

V2 (Vergeltungswaffe 2) 35–6, 37
velocity 89–90, 91
velocity fields 118–19
Venetian glass 193
Very, Frank 160, 161
volcanic eruptions 151
vortex 131–46

Walker, 'Boomerang' Gilbert 103–4, 144
Walker circulation 104, 105–6, 106–7
Waltham, David 26
water
  boiling 64–5
  steam 64
water molecules 25–6
Watt, James 167, 169
wavelength of radiation 61–2
wavelengths 156–7, 213
  of the Earth 62–3
  in metres/nanometres 62
  sunlight 63
waves 141
weather xiv–xv, 6, 8, 213
  British Isles 93–4, 99
  chaotic nature of 128
  climate and 93, 147–8
  definition 93, 147–8
  European 98, 99
  local causes 47
  predictions of 94
  in situ development 46
  strategic points 103, 144

wind and 43–4, 46, 47
  see also jet stream
weather forecasts 109–30
  accuracy of 129, 130
  digital 124–5
  by hand 124
  inaccuracy of 128, 130
  opposition to 115
  quantitative and qualitative changes 122–3
  Robert FitzRoy 112–16
  witchcraft and 114–15
weather fronts 121, 122–3
weather observation networks 111, 121
Weather Prediction by Numerical Process (Richardson) 123–4
weather systems 98–9
Wernher, von Braun 34–5
west-east motion 90
westerly wind 85, 95, 136, 213
Whipple, Francis John Welsh 132–3
wildfire season 182
Willughby, Francis 77
wind shear 136
wind speeds 44
wind(s) 43–54
  definition 44
  Etesians 73, 74, 76
  explanation of 79–80
  flows of 45–6, 51
  Halley's map of global wind patterns 79, 80
  weather and 43–4, 46, 47
  see also trade winds
winter 133–4
Witchcraft Act (1735) 114
Wolverhampton launch site 3

Woollings, Tim 97–8
work 88
World Meteorological
    Organization 148, 180

Yorke, James 128

Zeus 6–7
zonal symmetry 142
zonally symmetric atmosphere
    140